Managing to Communicate

Managing to Communicate

Using Telecommunications
for Increased Business Efficiency

Martin P. Clark
Deutsche Gesellschaft für Netzwerkdienste
Deutsche Bank Group Germany

JOHN WILEY & SONS
Chichester • New York • Brisbane • Toronto • Singapore

Copyright © 1994 by John Wiley & Sons Ltd.
Baffins Lane, Chichester,
West Sussex PO19 1UD, England
National Chichester (0243) 779777
International +44 243 779777

Other Wiley Editorial Offices

John Wiley & Sons, Inc., 605 Third Avenue,
New York, NY 10158–0012, USA

Jacaranda Wiley Ltd, 33 Park Road, Milton,
Queensland 4064, Australia

John Wiley & Sons, (Canada) Ltd, 22 Worcester Road,
Rexdale, Ontario M9W 1L1, Canada

John Wiley & Sons (SEA) Pte Ltd, 37 Jalan Pemimpin 05–04,
Block B, Union Industrial Building, Singapore 2057

Library of Congress Cataloging-in-Publication Data

Clark, Martin P.
    Managing to communicate: using telecommunications for increased business efficiency/
M. P. Clark.
    p. cm.
    Includes index.
    ISBN 0 471 94188 3
    1. Management—Communication systems—Technological innovations. 2. Business—
Communication systems—Technological innovations. 3. Telecommunication—Management.
4. Computer networks. I. Title HD30. 335. C55 1994
658. 4′ 5—dc20   93-30453
                                                                                                            CIP

British Library Cataloguing in Publication Data

A catalogue record for this book is available from the British Library

ISBN 0 471 94188 3

Typeset in 10/12pt Times from author's disks by Keytec Typesetting Ltd, Bridport, Dorset
Printed in Great Britain by Biddles Ltd, Guildford, Surrey

Contents

4 Networks for Computers and Electronic Equipment 61

5 Message Networking Services 87

6 The Company's Backbone Telecom Network   107

7 International Networking   129

8 Communicating with Confidence   143

Preface

When asked to cite the greatest challenges facing them, the answers of many companies include their need to improve communication. Often this is seen as the most pressing need. It is thus surprising that so little effort seems to be applied to understanding communications needs and to structuring communication paths. What I believe is needed is a formal approach for linking the human problems of communication with the potential technological solutions already offered by telecommunications.

Formal methods for defining processes have evolved from the computer and software industry, but in my experience these do not go far enough in helping with the choice of communication medium. This is a pity, for the opportunities abound for massive cost reduction and improved business effectiveness through better communication. The current costs of poor communication are much more extensive than the mere costs of technology and phone bills. Poor communication should also be costed in terms of the effects of conflicting information, inaccessible or mislaid information, distorted or overdue information.

Generally the measures needed for improving communication are quite simple. Reduction of multiple-step communications processes to single-step ones improves speed and reduces cost. Eradication of duplicate paths removes confusion and cost. Simply getting the choice of communications technology right can instil confidence and speed in the business process—a reinforcement of good practice.

The foundation of a new science I would not claim this book to be. But I hope that it brings a new focus to company communications needs and structures, and to the capabilities of technology. I hope you will find possibility and pragmatism amongst its pages.

Martin Clark
Frankfurt, Germany
July 1993

Summary

In recent years, telecommunications technologies and methods have advanced and diversified enormously—bringing cheaper and more powerful new opportunities for improved communication. Meanwhile, general businessmen have been slow to take advantage of the full benefits, and many companies remain ineffective and inefficient communicators.

This book sets out to redress this situation by laying down a framework for management of telecommunications technology for maximum business benefit. It describes the most important aspects of the technologies available to companies, explaining how to match these against critical business communications requirements and realise the benefits while simultaneously minimising costs. In short, it is a book about what needs to be achieved in communication, why and how.

The book is aimed at managers responsible for company telecommunications, but is not just restricted simply to 'telecommunications managers' and technologists—a prime aim is to raise the expectations of non-technologically-minded senior business managers for what can and should be achieved. Business is too dependent on its communication for the subject to be left to chance.

Many managers five years ago thought of PCs only as word processors for secretaries. Nowadays the personal phone and the laptop computer are a major part of business life, as are the electronic mail and electronic data interchange methods of communication that go with them. The ability to assimilate these, and other new techniques, may be a critical influence on the company's success.

The book's approach is one of simplicity and pragmatism, concentrating on telecom management objectives, style and technique—but always based on a thorough understanding of technology and how it can be made to serve business.

1

The Company and its Communications Needs

By improving their communications, companies stand to make significant improvements in the efficiency and effectiveness of their core business processes. This chapter challenges companies to understand the benefits of improved communication, to develop more clearly defined business communications needs, and, by so doing, achieve better Value for Money in their use of communications technology.

THE COST AND BENEFIT OF COMMUNICATION

A typical major trading company spends around 0.2–0.5% of its annual revenue on telecommunications services and equipment. This might be 25% of total annual expenditure on Information Technology (IT), but this proportion is likely to be growing. Thus a corporation with annual turnover of $200 million, could be expecting a telecommunications bill in the order of $1M per annum out of a total IT bill around $4M. A small price to pay—just one of the overheads of doing business you might say. But my own thoughts are more inquisitive:

- Are you paying too much for what you need? or for what you get? After all, a $1M increase on $20M profits would be welcome.

- Conversely, should you spend even more money—say $2M or $4M—to achieve even better value? Would further investment create further savings or revenue elsewhere?

- What is the value of good communication? And do the obvious telecom charges represent all of the costs?

These questions are hard to answer. It requires judgement and experience to assess the correlation between a company's effectiveness in communication and the effect of this on its overall annual performance. This is because individual communications do not usually contribute direct financial benefits. But our questions need to be a main guiding influence on the telecom manager and his corporate management.

The first question is perhaps the easiest to tackle. Getting the same or equivalent service for less money will clearly make a direct contribution to profits. In consequence, the cost-cutting priority is rarely overlooked by corporate management. Unfortunately it is too often the only objective given to a corporate communications manager—a consequence of what you might call 'incremental management': 'Do what you did last year but do it 10% better'.

The expectation of corporate management ought to be greater—and to have higher expectation they need to understand the issues and to get involved. Improving upon last year's networks merely by some arbitrary cost percentage assumes that what you did last year was right. But did you think about it? There are some crucial questions to be asked about business communication efficiency—for the enormous advances being made in technology potentially offer new opportunities for radical improvement.

A more challenging approach should be used: 'What causes us to need to spend anything like this much on our established communications? What new developments offer us potential for improvement in our business networking and our contact with customers? What is the value of good communication, and the cost of poor communication?'

It is now widely accepted that telecommunications are an essential foundation for business and national economic growth. But such a philosophy, while encouraging for telecom professionals, does not help in determining what communications tools are needed and how much should be spent on them.

For some people it may seem self-evident that each company employee needs an office, a desk, a chair, a telephone, and maybe a PC. Maybe in some companies this is true, but why should a telephone be needed in a supermarket? If the tills are connected to a data network—allowing automatic stock-taking and financial reporting to headquarters, as well as providing a simple electronic mail service for company notices—then what function will the telephone have? Perhaps to distract the employees from serving the customers? Or to allow the shelf-stackers to call their distant relations?

It is not for me to presume there is not a good reason for a phone in a supermarket, and far be it from me to suggest that company phones do not have important social functions as well. But such policy should follow decision rather than just come into being.

Figure 1.1 The intended use of the supermarket phone?

CORPORATE COMMUNICATIONS AS A BUSINESS

Telecommunication in any large company needs to be viewed and run as a business in its own right—with commercially guided objectives, and a clear expectation of getting value for money. The internal telecom department should be expected to deliver the services that the communicating customers need—at a price which is competitive alongside the alternatives. They need to be responsive to change in the types of services required and also to fluctuations in demand.

Extra money spent on telecommunications can be money well spent. A much quoted example is that of the American hospital supply company that equipped its customers (hospital wards) with computer terminals and network connections. When the ward needed more supplies of bandages, antiseptic, or whatever, it was made very easy for the ward staff to reorder. Competing suppliers who relied on telephone orders and letters sent by post started to lose out. In this case, improved communication helped take the company nearer to its market and so lock-in its customers.

Another example is in the introduction of EDI (*Electronic Data Interchange*). This is the name given to the method of computer networking which allows different companies' computers to talk to one another (for orders, invoices, information updates, etc). EDI is used by some supermarket chains to lay-off orders on their food suppliers. It benefits the supermarkets by allowing them to stock their shelves on a *just-in-time* (JIT) basis— more frequent orders and deliveries enable actual shelf stock levels to be kept much lower, yet still capable of meeting day-to-day fluctuations in demand. In the case of some of the stores in the '7-Eleven' foodstore chain in Japan, just-in-time delivery is conducted three times per day. From a very small shop a very high turnover is thus achieved.

EDI is also used heavily in the travel industry. The American Airlines ticket reservation system (which was made available for other airlines' use as well) is said to be worth more as a separate business in its own right than the aircraft operations part of the company. The system allows travel agents to check seat availability and make direct reservations on behalf of their clients. The key to success was widespread availability of the system amongst agents.

DETERMINING VALUE FOR MONEY

The central dilemma for corporate telecom managers in determining how much money to spend is how to associate a value with good communications, and by so doing create a framework on which to judge the value of technological investment. It is clearly not an exact science, but this should not become an excuse for avoiding it.

Putting a value on communication is most easily done by calculating the cost of some business process as a whole, rather than merely by concentrating on the totalled cost of technology and its use—the network, the phonecalls and the individual data transmissions. Allocating costs to the business process allows the company to think on its own terms—'Cost of taking an order' etc—and gives the telecom manager a clearer objective for how he can work to help the business. He can then much more easily focus on what he must achieve in technological terms.

Consider the following examples of business processes, the success of which depends very much upon the speed and quality of overall communication. Consider in particular how the process as a whole could be streamlined or shortened, and then reflect upon whether the right technologies are available to the right individuals in the chain.

How much does it cost the company to take an order?

- How many orders are wrong, and why?
- How much does it cost to correct them?
- Would a change in the process, and perhaps in the technology make the function more effective and/or reduce the cost?

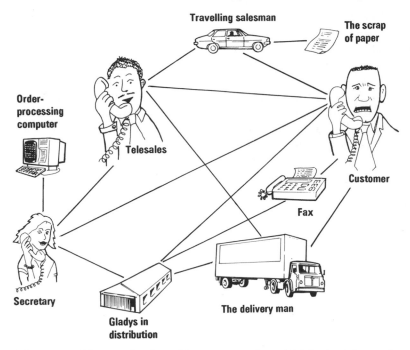

Figure 1.2 The usual order-taking process—but which is the right way?

How much does it cost the company to invoice and get its payment for a delivery?

- How many invoices are wrong, and why?

- Do invoices reflect the orders, or the deliveries? Which should they reflect?

- Why don't deliveries match orders?

- How could the invoice be moved closer in time and place to the delivery, so that monies are received earlier?

How does the company monitor its customer service?

- How many complaints are received?
- Does the customer know how to complain?
- What is the cost of poor service? or of lost business?

By redefining communications costs in these terms—including the (non-technological) cost consequences of poor communication (eg the human time taken in correcting errors, or the lost business resulting from poor service)—we can create a much greater opportunity for radical improvement, and improved technological investment (if appropriate).

I advocate, for example, delegating the job and cost of running the internal mail system and of organising express mail courier services to the telecom manager. In this way I believe you create incentive and potential funding for some significant communications network improvements. But having said this, I do not advocate technology for its own sake. The company's communication networks, in whatever form they take (be they electronic, verbal or whatever), are there to support the core business of the company, and this must clearly come first. A short letter can be a most effective and efficient communication. The day's delay before delivery may not be a problem.

THE TECHNOLOGIES AVAILABLE

What technologies are available to help us ? Many are so familiar that they are taken for granted. They are easy to understand and barely a thought goes into their use:

- Face-to-face conversation
- Telephone
- Fax (facsimile, or telefax)
- Telex
- Telegram
- Television and video
- Radio
- Radiopager
- Computer networks.

Other new technologies are also appearing and gaining rapid acceptance:

- Cellular radio (portable and carphones)
- Voicemail
- Videoconferencing.

These too are falling into everyday life. Suppliers have paid such attention during the development of products in making them easy to understand and use that no second thought goes into how and when to use them. The rapid emergence of fax in the early 1980s came about because of its simplicity and speed of use. It created a new market for itself by enabling ordinary people to discuss complicated matters easily over a course of only a few hours — sharing draft wordings of papers and contracts, and diagrams of new concepts.

But maybe the lack of thought in how we use communications methods, and which one to use, is itself a problem. Have you ever stopped to think how efficient we really are in telecommunication? Are the phonecalls we make always necessary and well-timed? Are the messages we send by fax always as urgent as we make ourselves believe? And do they reach their recipients as urgently as we would like them to? If not, is this really the best method to use? How much do we spend on fax? What are the alternatives?

We listed above a set of familiar methods of telecommunication — all successful ones. There is a long list of other, and alternative, methods to those above about which you may know little. Many of them are initially difficult to understand. Many are poorly merchandised — not instantly recognisable as 'business solutions'. The question is how potential users can become more familiar with them, and how suppliers can become better at merchandising and selling them.

The telecom industry still has a long way to go in product merchandising. Too many products still take a considerable amount of specialist time to understand and to assimilate into effective business use. You might, for example, ask 'What is an X25, and why might I want one?' Unfortunately, the reaction you are likely to get from a telecom expert is only a smug grin and a smirk — 'who is this idiot?'. For X25 is not a particular item of equipment, but instead a complicated technical specification — one which defines a particular type of communication method by which computers may 'talk' to each other across data networks. X25 might be employed, for example, as part of the 'language' used between electronic tills in individual supermarket locations and the central data communications centre.

Why should the average corporate manager be expected to know this? Why do the suppliers boast so many products but not tell you clearly how to use them in terms that you can understand? Why indeed?

It isn't even clear to me why telecom managers should be expected to know about them. New technologies are appearing at a fantastic rate, and the jargon is rapidly increasing in consequence. No wonder, then, that executives and indeed whole companies retain their technology-phobia. The telecom supplier loses out from this situation through lost sales potential. But the sad fact is that the customer is also put at risk — from the competitor who can assimilate critical technologies first.

What is lacking is a good understanding by the suppliers of their customers' real core business needs. The customers meanwhile may suffer near complete technological ignorance, and even phobia — put off by many of the complicated concepts, language and jargon. To bridge this gap, and create better cross-fertilisation of knowledge and ideas from both sides, I suggest that what is needed is a new and much simpler language of debate. It needs to be possible for even the least-technologically-minded manager to describe his communication needs in his own business terms, and for the telecom manager to interpret this in terms of the technologies which he may offer as solutions.

In the remainder of this chapter we discuss the main attributes which distinguish the various requirements of business communication, and the possibilities offered by technology. By understanding business needs in these more simple terms, we strive to make every business manager more discriminating in his choice of telecom technologies for particular purposes.

TECHNOLOGY CHARACTERISTICS

What sort of attributes and characteristics should be considered when developing communication networks and selecting telecommunications technologies? I suggest the following:

Attribute required	Particular considerations, for example:
Volume and frequency	How much needs to be said or sent, when, and how often?
Image/quality level	What quality of print is required? How much 'frying bacon noise' can be tolerated on the telephone?
Information accuracy	Does a lot of detailed specification or order information need to be transmitted? What is the shortest route and best carrier for such information?
Reliability/assurance of delivery	Is it critical that no messages are lost? Can they be repeated if necessary?
Confirmation of receipt	Is a confirmation of receipt needed? And on what timescale? Would it still be needed if receipt was assured every time? (ie is it just for peace of mind?)
Answer/response needed	Is the reply what is needed? But why is this prompt for reply necessary?
Time available for delivery	Does it really need to be immediate? Can it wait a few hours/days?
Broadcast	Is the same information (price updates, race results, product updates) to be sent to a number of people or locations? Could enquiries be minimised if *all* the information were sent to *everyone*?

Collection of information	Does information have to be collected from a large number of sources? (eg supermarket sales results and takings)?
Human contact	Is personal contact required? Is there an element of negotiation?
Record of communication	Is a 'hard copy' needed? Does the order have to be filed? Is the signature important?
Security of information	Is the information 'for your eyes only'?

Only some of the attributes listed above need be known to enable us to make the correct *choice of technology* for a given application. (For example, the need for immediate human contact may demand the use of the telephone, but the telephone cannot produce hard copy.) The remaining attributes (particularly the volume and frequency) help to determine *how best to realise the network* itself. (For example, should an external service supplier be used or should an internal solution be developed, and what capacity is needed?)

When considered in a discriminating manner, the attributes clearly indicate which technology is most appropriate for application. The examples below illustrate some simple differences:

Attribute required	Example technology comparison
Volume and frequency	Facsimile is quick for a short document, but the ordinary letter post is cheaper or an electronic mail system is both quicker and cheaper for very long ones.
Image/quality level	Facsimile is quick, but final print quality can be poor. Post or electronic mail can produce higher quality printed results.
Information accuracy	Taking complicated orders over the telephone can lead to errors. A form of some kind is better (either in computer or paper format).
Reliability/assurance of delivery	Fax messages and ordinary post can be lost, even just in the internal mail. A registered letter or phonecall is safer.
Confirmation of receipt	Telex tells you for certain who received the message and when. Fax recipient identifications are unreliable.
Answer/response needed	A telephone call can prompt immediate response. Letters often go unanswered.
Time available for delivery	A telephone call is immediate, but only if the person you are calling is there.

Broadcast	Satellite or radio broadcast is an excellent medium for broadcast. The repetition of telephone or fax calls is by contrast very tedious.
Collection of information	There are four options for receiving multiple reports or enquiries: (i) big system handling multiple calls; (ii) queue calls and answer in turn; (iii) log calls and call-back when ready; (iv) have a strict timetable for call-in.
Human contact	Meetings, phonecalls, and video or voicemail messages provide differing degrees of human interaction.
Record of communication	A phonecall has little record, while a fax provides a hard copy for both sender and receiver.
Security	Personally carried messages are safe; many other methods are prone to 'tapping'.

At a much more advanced level, the same attributes can be even more discriminating than so far illustrated. As an example of this, we consider in detail only one of the possible required attributes—the need for a response. In the next section we try to be more critical about how quickly we need the response, and how much we are prepared to pay for faster service.

THE COST OF QUICK RESPONSE

Figure 1.3 and the associated table below illustrate the proposition that the quicker we need a response to a message then the more we must be willing to pay—conversely, the longer we are willing to wait, then the more money we can save.

Figure 1.3 plots, for different types of telecommunications technology, the relative cost of sending a message and receiving its reply against the most likely (ie realistic and repeatable) response time. The diagram assumes both message and reply are two pages of A4 paper in length, and that this equates to five minutes of conversation. The costs include the usage costs of sending the message plus an apportioned part of the overhead costs (eg equipment rental or depreciation costs).

Technology	Cost	Likely response time
Cellular phone	5 mins × 50p per minute.	Immediate, if the phone is always carried.

Radio pager plus phone	5 mins × 25p per minute (public callbox telephone) plus 35p share of monthly pager rental (£10 per month).	Typically it takes half an hour for the called individual to break from his immediate engagement and find a phone.
Telephone	5 mins × 20p per minute (normal phone call tariff) plus 20p share of monthly rental.	Immediate response if the called individual is at his desk, but in half of cases maybe he calls back later, in response to a left message.
Voicemail	0.75 minute telephone use × 4 × 20p per minute plus 20p share of monthly voicebox rental or equipment depreciation (£6 per month). (Voicemail messages tend to be shorter and more to the point than telephone conversations. At least four calls are needed to leave and pick up both messages.)	Workgroups which employ voicemail as their main means of communication typically achieve message response within the working day. It takes 2–4 hours for the first message to be picked up, and then a further 2–4 hours before the initial sender once again checks his mailbox.
Electronic mail (email)	Based on the use of a public electronic mail service costs are assumed similar to those of voicemail. The costs of a private company email system might be much higher or lower than this dependent upon the depreciation charges for computer equipment and software, and the overall level of usage.	Similar response time to voicemail amongst workgroups who use it—typically within the working day.
Facsimile (fax)	4 pages takes 4 minutes telephone time, so 80p telephone charges, plus 15p for special thermal fax paper and 5p equipment depreciation plus 20p share of monthly line rental.	Rarely immediate since it usually takes the receiver until he next looks through his post to realise the message's receipt.
Post	2 × 25p postal charges plus 10p for paper and typing time.	Post is nowadays rarely treated urgently, 3–5 working day minimum for reply.

The general trend of our hypothesis seems generally to be borne out in the model—the quicker we need response, the more we must pay. But there are some shocks, such as, no doubt, the very high costs and response time attributed to fax. I justify my reasons later in the book (Chapter 5). To some extent, of course, the results are due to the assumptions we make to start,

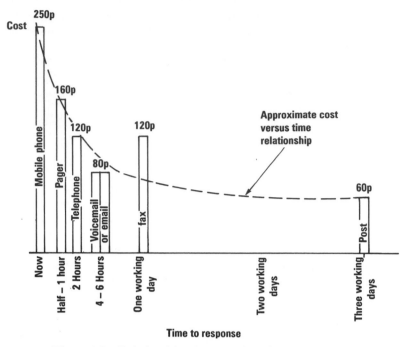

Figure 1.3   Relationship of cost against time to response

and upon certain subjective judgements. My assumptions may appear shocking at first, but I have aimed to be objective. If you are not satisfied with my judgements and costs, then of course you are free to derive your own.

BEING DISCRIMINATING IN DEFINING NEEDS

When defining business needs in terms of the previously discussed 'attributes', you need to be especially careful to distinguish between what actually needs to be communicated and how the business operates today. If you are not careful you will find you are reviewing the technology and upgrading it, rather than considering the underlying communication need. The example which follows attempts to illustrate this point.

The UK bookmakers (ie betting shops—Coral, Mecca and William Hill) revolutionised their businesses in the 1980s by the introduction of satellite technology. Nowadays they broadcast all possible sports results to all their shops (Figure 1.4), irrespective of whether the shop has taken bets on a particular sport or not. The individual shops are equipped with personal computers which sort the information for their own needs.

The previous system was much less capable. Live horse racing commentary from various venues, interspersed with the results from all locations, would be broadcast by private wire or radio to all shops. This was not only an expensive method of broadcast, but also left a need for a large number of enquiries about certain results to a central bureau.

Satellite makes the broadcasting much more efficient—one-way information flow bringing the spin-off benefit of live television action for the punters in each of the shops, from various locations simultaneously.

BUSINESS COMMUNICATION NEEDS

So down to business. What sort of communications needs do real businesses have? How can we express these in terms of our newly defined attributes?

This job is too important to be left entirely to the telecom manager, and should involve the attention of every business manager. The telecom manager cannot single-handedly be expected to understand the value of communication in every department or subsidiary company business.

Let's look at a few examples to illustrate how the approach could be applied. We consider in turn some possible communications needs of companies in the following business sectors: retailing, manufacturing, and trading.

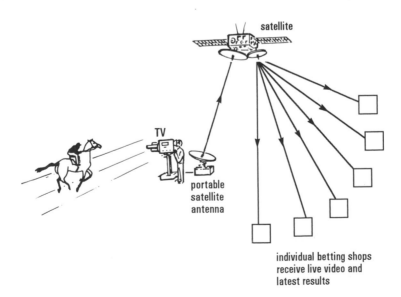

Figure 1.4   Satellite usage in the bookmaking business

Communications Needs in a Retailing Business

In this sector we consider the communications needs of a chain of retail fast-food outlets (say Burger King, McDonalds, Pizza Hut or some such), and describe their main communications needs in our new terms.

Unlike other restaurants, these chains rely on a rapid throughput of customers — typically aiming to see even the 'sit-down' customers leave within 10 minutes. The fast service is what the customer is after, and furthermore, is what enables the restaurant chain to make the best use of its limited space.

In order to achieve the high turn-around, a limited range of products are offered, but in various combinations to suit customer taste. The preparation of these products is done in the 'back of shop' to exacting logistical procedures.

Counter procedure is friendly and courteous, but also 'no-frills'. A considerable effort is put into 'point of sale' advertising in an attempt to increase the value of individual purchases — 'Would you like a Coke or a dessert, sir?'

The successful amongst these businesses have clear marketing images, and pursue high level advertising to create and maintain customer awareness. Geographic positioning of the 'restaurant' is all-important in order to pick up passing business — shoppers, train travellers, etc. Awareness of trends in sales, even in the short term, allows the adjustment of product range, the honing of promotional campaigns, and the appropriate density and location of new restaurants.

To some extent individual restaurant managers and franchises are left to get on with their own business, but the headquarters has a very strong support role to play in providing the right image, products and promotion at the right time, also in supporting the logistics of the burger 'production line' — providing the buns and the burgers, and keeping a strict control on restaurant-output quality.

The basic communications needs of these companies include the following:

- Restaurant to headquarters (*HQ*). Communication of sales breakdown (nowadays by product, price and time of day at the very least). This communication is individual to each store. Sometimes each sale is reported individually as it happens. Sometimes a consolidated report is compiled and sent daily. The information is used both by marketing for business analysis and direction, and also by the distribution arm, which supplies the burgers and buns on a just-in-time (JIT) basis, to ensure freshness and minimum stock levels.

Need: collection of large volume of numerical information from multiple sites on at least a daily basis. (Could be met by using a dial-up computer data network used once per day from each store.)

- HQ to restaurants. Broadcast of advertising and product information, as well as price updates. The same information may be broadcast to all restaurants weekly. Another 'need' also considered by some companies has been the transmission of a specialist TV programme, designed both to entertain guests, and also to provide direct point of sale advertising. (Imagine a Coke advert showing behind the till as you purchase your burger).

Need: broadcast method. For some information once a week may suffice (by post perhaps?). For the TV application it may suffice to send a pre-recorded videotape once monthly, but maybe more 'immediacy' is needed (eg using a live TV channel, broadcast by satellite to all stores).

- Restaurants to regional managers. Individual restaurant performance is usually managed by a regional manager responsible for ensuring uniform quality, service and profitability. They may require simple regular (say, weekly) business reports and in addition require an 'informal' means of quick communication (maybe using voicemail—see Chapter 5).

Need: a simple, cheap and quick way for mobile regional managers to contact restaurant managers or be contacted by them.

- Complaints Handling. Fast food companies are particularly sensitive to ideas and complaints, since this is a relatively easy way to secure and increase business from existing customers. Food contamination is potentially damaging if not diagnosed and contained quickly.

Need: a simple, but quick method by which customers may lodge complaints or suggestions. (Implement a call-in or write-in complaints bureau?)

Communications in a Manufacturing Business

Manufacturing enterprises usually comprise only a small number of sites, but often involve a large number of people. Day-to-day external communications with both customers and suppliers are demanded by the practice of just-in-time stock management. Customers do not wish to receive and pay for finished goods before they need them, and the manufacturers themselves similarly do not wish to order raw materials from their own suppliers before they need them.

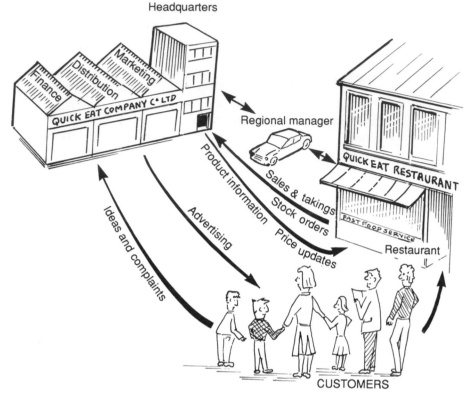

Figure 1.5   Main communication needs in a fast food company

On-site, sophisticated computers usually control most aspects of the design and manufacturing process, and good communication between the various departments will be needed for maximum efficiency of the process as a whole. A break in production during times of maunfacturing may cost the company heavily.

The basic communication needs are:

- Customer to factory, and factory to supplier. Frequent ordering of finished goods according to short term retail stock requirements, and invoicing for delivered goods.

  Need: some form of electronic data interchange (EDI).

- Internal factory communication. For computer control of machinery and inter-departmental communication. Maybe also, some emergency communication measures are necessary for safety reasons or for quick paging of specialist technical maintenance staff.

Need: high capacity and very reliable on-site computer network, and an effective and reliable means of personnel paging and communication.

Communications in a Trading or Brokerage Company

Trading and brokerage companies, more acutely than other types of companies, rely upon instantaneous information updates about the market in which they are operating. Perhaps more clearly than other companies, these are truly information-driven companies. These companies are generally highly computerised, and may also depend heavily upon electronic means of trading and communication.

- Information collection. Companies like Reuters, Dow-Jones and Telerate have developed entire businesses out of supplying information to financial broking houses. Statistics in many different forms are presented. But, while these are largely electronic, in other traded markets the market making and information collection may be entirely over the phone.

  Need: reliable source of information, with accurate and timely receipt (possibly achieved by means of a data network — see Chapter 4).

- Information analysis. Internal office systems may allow traders to refer to information provided by internal analysts on demand, and will cater for the ordering/invoicing support for the trading.

- Communication to clients and suppliers or other brokers. For taking and laying off orders.

  Need: for some legal record and execution of the transaction. Some traded markets use telex, others have progressed to fax, while the most sophisticated and fast moving markets use electronic exchanges.

TRANSLATING NEEDS TO NETWORKS AND TECHNOLOGIES

Having assessed in business terms what we are actually trying to communicate, next comes the task of determining which technology fits best. This requires a knowledge of the various technologies and their capabilities according to our attribute list.

Sometimes, individual business requirements each justify a separate network solution (of whatever technological type) tailored specifically to their individual needs. This is the easiest to manage, since the service can be provided or taken away according to the needs of a single customer. Monitoring of usage is easily attributed, cost allocation is straightforward, and problems can be relatively easily addressed according to a single set of objectives.

Unfortunately, however, networks usually have to be shared between a number of different types of user—with their conflicts of interests, requirements and desires. Communications network management in this case becomes more than a case of assisting individual business 'customers' to find communications solutions; it also requires a considerable amount of compromise—juggling, persuasion and refereeing.

Quality and usage monitoring of networks are often hard to assess, and service to individual types of users may vary, so that while some are happy, others are not so pleased. Cost allocation is always problematic. The decision of one 'customer' to withdraw from use of a network can, in some cases, load so much cost on other users that they also are forced to withdraw their use—even though the company may have already sunk all the capital costs.

Supplier management is always a critical determinator of costs and service levels, while technical specification 'standards' are critical in ensuring the ability of different equipment throughout a company to be able to intercommunicate.

Assessing needs, evaluating and implementing technologies and managing communications services is the task of the telecom manager. In the next chapter we suggest some management methodologies to help him.

2
The Communications Manager

This chapter describes a number of methodologies for company telecommunications management. The methodologies described, and the advice that goes with them, are intended to highlight the challenges and to help to overcome some of the obstacles to good telecommunications management.

COMMUNICATIONS MANAGEMENT DIFFICULTIES

The factors which contribute to good telecommunications management are:

(i)   Knowing what equipment is out in the company, who is using it, and what they are using it for.

(ii)   Knowing how to treat suppliers so that they give maximum support at the best price, but still feel they get a good deal.

(iii)   Coping effectively with monopoly suppliers.

(iv)   Keeping costs and quality at optimum levels.

(v)   Keeping knowledge and plans up-to-date with the latest capabilities of technology and the communications services available on the public market.

(vi)   Maintaining a workable policy stance with respect to technical communications standards, so that while incompatible equipment does not flourish, innovation is not stifled.

(vii) Knowing when to kill in-house telecom networks which have become white elephants.

In what follows we will tackle each of these areas, but will first consider whether this function does indeed require a communications manager.

DOES A COMPANY NEED A TELECOMMUNICATIONS MANAGER?

Given a typical company expenditure on telecommunications of around 0.5% of revenue then we might propose that 1 in 200 people should be allocated for telecom management. The corollary would be that any company of more than about 200 people would have a telecom manager.

But so much for theories. Communication in its wider sense is too important an issue to be left in the hands of only a few individuals. It is instead an issue to be addressed by every business manager, and costs much more than the money spent on telecom technology. The human costs of communication, and mis-communication, are enormous. In short, each manager owes it to his department, to his 'customers' and 'suppliers' to be his own communications manager.

In support, a technology-oriented department is helpful in assessing the capabilities of technology to support business needs, and also in realising solutions. But the danger of full-time technologist departments can be their over-preoccupation with creating and tending to technology. For example, too many internal telephone networks continue to exist in some of the largest companies despite the fact that using the public network may nowadays be a cheaper solution. Meanwhile the remaining company management does not concern itself with communication, it just takes it for granted, and anyway 'that's the communications manager's job'.

I am personally in favour of a new type of business appointment—the 'business communications manager'—a person developed from the traditional type of telecom (technology) manager, but more business-minded (Figure 2.1). Working closely with personnel, and to some extent business planning, they need to be the conscience of the company—continually challenging it to improve the basic patterns of communication and business process. And the challenge should not stop there. As traditional 'telecom managers' they should also be answerable for the delivery of technology solutions of the best value-for-money possible. Fluctuation in the size of the telecom department over time—to adjust for changing balances of internally and externally managed solutions—may be a necessary part of achieving this.

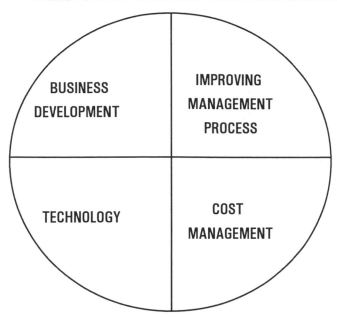

Figure 2.1   The role of the communication manager

WHERE TO START—ENLIST HELP FROM YOUR MAIN SUPPLIERS

It may seem obvious to suggest that a good starting point would be a complete inventory of existing telecom equipment in use by the company, a knowledge of the costs and the assurance that cost allocation (as appropriate) was being carried out fairly. The problem is that the inventory (and thus the costs) are often very difficult both to assemble and to maintain.

Because the unit cost of a carphone, an extra telephone line or a new fax machine is relatively small, it is often within the budget authority of many individuals within a company. Thus small orders may be being generated all the time, and billed direct to departments. At the same time, changes in personnel may mean that redundant services fail to get cancelled promptly. They may even go unnoticed on subsequent bills—particularly if they appear on the same high value invoice as other more-used services.

One solution to the problem is for senior management to demand that all services are ordered via a special telecom department. And this works very well in practice, provided the end business user perceives an advantage from dealing with the telecom department as a middle man rather than direct with the telecom supplier. The usual incentive for achieving advantage is a discounted price negotiated by the telecom department, or a better service.

Then the proposition is 'deal direct with the supplier if you must, but it is cheaper or easier to work with the telecom department'. In addition, of course, there's an internal ally for the business user to help solve any difficulties he has with the supplier.

A further refinement of the arrangement, which avoids the cost of a large telecom department just passing through orders, is to ask business managers to deal directly with a nominated individual at the supplier. You can then later ask this individual to produce an inventory of all sales. This enables you to track all new item orders. You save effort, and anyway it is generally a lot easier for a supplier to collate all incoming orders than it might be for a telecom manager to track outgoing ones.

To set up the arrangement, re-negotiate supply contracts for your main expenditure items (private network, cellular radio supplier, etc). Start the process by requesting all existing suppliers to supply an inventory of currently supplied items. You may be surprised to discover how many different cellular radio suppliers are used by the company. Simple rational-isation to a single supplier can generate a worthwhile discount (maybe 40% or more).

So the steps in sequence:

(1) Ask existing suppliers for an inventory of currently billed items.

(2) Tender the entire business to all available suppliers, demanding that a discount will be possible for *all* current users (otherwise there will be no incentive for all users to change supplier).

(3) Make sure the best-price supplier is also willing to process incoming orders at a single point (or at least collate a full sales report).

(4) Make sure that the supplier is also willing to bill your company in a manner appropriate to your needs (some require separate bills for separate business functions, while others with centralised accounting departments are able to recharge costs against individual accounts *).

(5) Finalise a supplier contract—for a two or three year period (the period length should be sufficiently long to be attractive to the supplier, and undisruptive for your 'customers' but also give you flexibility for change in a short enough timescale).

---

*Suppliers to large conglomerate companies can find it difficult to recognise all their incoming orders, particularly where individual subsidiaries of the conglomerate trade under their own names. This can be a real problem for a company rapidly acquiring or divesting of businesses, or one with subsidiaries which often change their names. My advice is to ensure, with senior management support, that an agreed convention of single company name and account number is used for all ordering by all subsidiaries. This may already be second nature to companies with strong purchasing departments.

(6) If your demand for the given service is growing, you should negotiate on the basis of anticipated rather than current volume, but you should not allow the supplier to impose any penalty clauses for failing to reach the target. The problem with these is that it may not be obvious which department should pay the penalty. (A department which has exceeded its forecast will feel aggrieved at being asked for a backpayment.) Instead, you should structure the contract to allow mutual price renegotiation (say at 6 or 12 month intervals). No matter how the supplier might complain if you do not reach forecast purchase numbers, you can rest assured that his unit price makes him a profit contribution on every individual item sold (though the margin might be low, and the overall profit he might have expected will obviously be affected).

WHAT TO EXPECT FROM YOUR SUPPLIERS

Have reasonable expectations. Choose the best suppliers and always remember that your own success depends upon them. Negotiate hard, and ensure you always get the best price. Make sure you are not paying more than your competitors. But always be fair. Strike the 'win–win' deal—of benefit to both you and your supplier. He too needs to feel he has struck a good deal if he is to give the service and support that you expect.

You can skin the occasional supplier with a 'win–lose' deal, but remember that he may not want to come back. A long term reputation for such deals will turn the supply market against you.

What can you expect? The simple answer is 'the best value that the market can offer at the time'. You achieve this by being well-informed about the market. You need to know what the alternatives are and how much they cost, for only by comparison can you determine the best value for money. Never let yourself believe that you have no choice.

You can expect suppliers to keep you up-to-date about their new and changing products and prices. But more than just that, you should ask them to be explained in the context of how they might be useful to your company. Your supplier owes you the duty of understanding your business, and striving to improve his support for it. For just as your company's success depends upon your supplier, so your supplier's future business depends upon you.

Most attention from a supplier is achieved when there is potential new business for him, or the risk of losing his current business. These opportunities are therefore valuable, and should be exploited to their maximum. At these points in time, a tendering competition involving other potential suppliers helps to keep the supplier on his toes, and gives you an invaluable 'snapshot' on the exact state of the market and prices at that time. But do not attempt such reviews too often, for they are expensive in time both for you and your supplier. And remember—too many idle threats, or too many price requests of suppliers at times when you have no serious intention of giving them business, are only harmful to your credibility, so avoid them.

NEGOTIATION OF SUPPLY CONTRACTS

Always recognise and understand your alternatives. If you believe you have none, then presumably you are willing to pay any price!

Negotiation is always most easily conducted with suppliers competing against one another. It is then just a case of refereeing to determine which is the best offer. But make sure the process is given enough time.

The process can start either from a clearly specified need defined by the customer or can be triggered by interest caused by an offer made cold by one of the suppliers. The aim of the negotiation process will be to confirm the validity of the customer's need as specified and then determine the best value solution. Avoid (if possible) requirement specifications which are only really descriptions of the current technological solution (or worse still, a particular supplier's product) for this tends to exclude some of the alternative options too early, and enables the existing suppliers to bid a high price, knowing that the competition cannot match them on technology.

The negotiation itself (Figure 2.2) is only a process of offer and counter-offer made by each of the competing suppliers in turn. As each offer is made, discuss the strong and weak points of the offer with the supplier who has made it, and ask him to improve upon it. Also let him know the strong points of other suppliers' offers—where he is behind—and ask him to respond.

Your comments need to be truthful and relatively specific, but you should not (and need not) breach any commercial confidentiality: 'Overall you are about 30% more expensive' or 'Your maintenance price is competitive, but on the price of purchase you are in last position—50% out of line' or 'Your competitor has some interesting ideas for co-marketing—can you offer anything?' Through such comments you need to make sure that all the suppliers are absolutely clear what your main requirement is, and what your basis for decision making will be.

Let the process run its full course—until the suppliers appear able to offer no more. You may find at this stage that there is only one serious contender left in the running, but if not then you need to make a decision between the suppliers who still remain. The decision making may well be easy and acceptable even to the losers, particularly if they have felt informed about where they had to improve (but failed to). They will have gained the

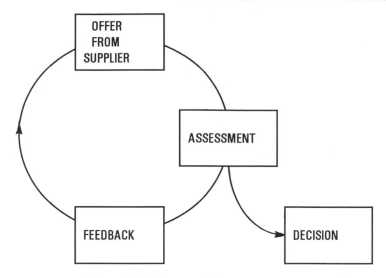

Figure 2.2   The negotiation process

knowledge as to how to be better prepared next time—and this should be a benefit to both sides.

If the loser complains about losing, then remind him of his offer's weaknesses and ask him to bid again—maybe you had misjudged what more could be offered.

The process should be self-regulating. Suppliers usually only make offers which are good business for them, even though they may appear to have discounted heavily from their opening negotiation position. The customer chooses what he believes is the best value, and by so doing influences what the market will offer in future.

However, particularly large or influential customers may need to be a little careful about other influences impacting some suppliers' offers. When a customer dominates too much of a given supplier's capacity, it may be that the supplier cannot afford to lose the business. In this case he may bid a low price to retain the business, but straight away may have to start cost cutting to make it profitable. Service may drop off in consequence—a pity if the supply relationship has always been good, and the price not so critical to the customer. The old adage says 'always leave something in the deal for the next man.'

Finally, you may need to commit the agreed commercial terms to contract. Try to do this as fairly as possible. Remember the job of both lawyers (or equivalent) is merely to encapsulate and record the deal, in the spirit in which it was negotiated. It is possible to alter the deal quite substantially in the legal fine print, but this may damage the goodwill achieved during negotiation. If you were happy with the deal when negotiated openly, why do you need to seek underhand means of improving it?

DEALING WITH MONOPOLY SUPPLIERS

Because of their historic monopolies, the main suppliers of wide-area telecom networks, the PTTs (national post and telecom network companies), do not have a history of flexibility and willingness to negotiate.

Some PTTs have become much more commercially aware in recent years and, in consequence, much more customer-focused. In other countries, recent or imminent changes in national regulations will bring much more competition within the next few years. But, despite these beneficial influences, the incumbent operators are bound to remain market dominators (and perhaps inflexible) for some time. Indeed, for some outlying geographic areas, the monopoly supplier may remain a monopoly for ever— simply because other companies fail to challenge this business.

So how do we negotiate and deal with these companies? There are three guiding principles:

(1) Recognise that your business, particularly your future business, is important to the PTT.

(2) Be more inventive when seeking alternatives.

(3) Understand how PTTs work, and present your requests in pieces that their organisation can digest.

PTTs facing imminent competition are likely to be seeking a good reputation with their government—and so even the mere threat of competition can sometimes stir the PTT to better service. Your good opinion of them may, on its own, be influential. Helping them to gain a better credibility within your own industry might generate an even better return. If you have to resort to threats, an industry pressure group lobby or a telecom users lobby are the best alternatives. These help you to confirm your poor impressions and give strength in numbers.

Even in PTTs not outwardly facing competition, there are likely to be internal tensions and competitions within the PTT which can be exploited. For example, the product manager of a new network service may be seeking to make his mark—maybe at the expense of other services. Many PTTs historically were cumbersome organisations, with poor internal communication. This meant there were some inexplicable anomalies and, sometimes, directly competing services.

Non-cost-based pricing by the PTT can lead to some outrageous pricing anomalies, which also may be exploitable.

The inflexibility of the PTT as a whole, and the national regulatory situation may preclude the short term lower prices that you may deserve. But other gains are likely to be possible in the service level achieved—the friendly technician who always beats the standard delivery time for new circuits, or the fitter who will wire-in your extra telephone sockets for no extra charge.

In the longer term, unless the PTT continuously improves, then the case for change, or increased competition, grows against them. The telecommunications section of the recent GATT (United Nations General Agreement on Tariffs and Trade) agreement will see to that.

A big company can exert more immediate pressure on its own. For international companies, the monopoly of a PTT in one country might be the very reason why the company elects to build its new manufacturing plant somewhere else or, at the very least, why it chooses to create its main telecommunications hub or computer system elsewhere.

Some international corporations may be able to offer PTTs the opportunity to expand into other countries. After all, even the biggest PTT may have only a small proportion of a multinational conglomerate's current telecom business. Taking all the international business of their existing major corporate business customers would represent a huge opportunity for PTT growth.

Never ask for too much at once. It is much more effective to allow the PTT to see some quick and uncomplicated gains first, as the encouragement to bigger things. This helps your PTT counterpart deliver within his capability, and builds his influence to help you with the bigger task.

THE CORPORATE NETWORK

Success in creating a corporate network depends largely upon setting the right objectives. In my view, the corporate network should be thought of only as a concept—one comprising all possible communications means within and without the company (Figure 2.3).

The internal and external mail services should be considered part of the corporate network, as should the costs (time and travel) associated with regular management meetings and reporting procedures. These are all obvious candidates for potential change by alternative telecom technologies.

Other costs which could also be included are photocopying costs, secretaries' time and the costs of travelling. All of these can be impacted by better communication. Finally, in the future the whole cost of the office and its infrastructure may need to be included as 'communication costs', for the concept of 'teleworking' proposes that sophisticated telecommunications technologies can remove both the need for and the constraints of offices!

And what is the job of the communications manager?—To manage and improve this network.

What place, then, has the corporate telephone service? It has a binding influence upon the company—keeping various organisational parts in close touch, despite perhaps huge geographical and organisational barriers.

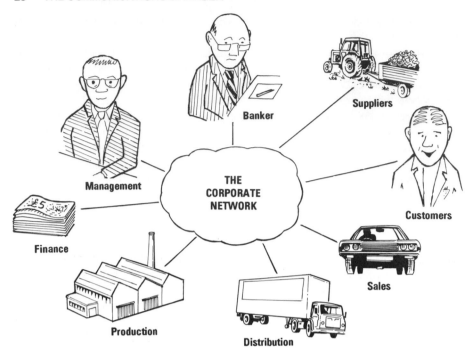

Figure 2.3   The corporate network

The company's private telephone network may or may not be critical to its telephone service. A private network may have a place in keeping down the cost of telephone usage—or for stimulating more communication at very marginal cost. It may also have a role in keeping customers closely bound in—using direct telephone or data connections. There may be even more benefits. But the reverse is also true: a private network which may once have made economic sense, may not anymore. Certainly the need for it will change over time and the network itself must be adjusted to accommodate this change.

Just like the product of any other business, a telecom network needs to be recognised as having a life cycle—a beginning and a period of growth, followed by a period of stability and maintenance, followed by its timely termination. Just because you always previously needed a network of a given type doesn't mean you need one now. A good telecom manager does not just compare costs of the old telecom network with a replacement one using the latest technology. He should be acutely aware of the changing needs, the cost of the alternative solutions and the justification for his chosen option.

Before choosing between alternative options and particularly before establishing private networks, make sure to understand the downsides and the risks of investment, as well as the benefits and savings. Telecom networking is a capital-intensive investment, and often requires a fairly long period to achieve full payback. Also be careful about add-on investments. In large networks, unit costs can actually increase, because the complexity of the 'spaghetti' of circuits gets beyond the capability of the management to find optimum routes through it.

There is nearly always the possibility of using a public telecom network service instead of a privately built network. And while public networks may appear in the long term to be more expensive, the benefits will be the reduced capital investment, the reduced technology risk, the shorter-term notice of termination and the reduced commitment to forward costs (manpower and depreciation).

The scale of economy that ought to be achievable by PTTs should enable them to price public network services much more competitively than most private networks could ever be. But, this said, there has been, and will continue to be, a place for private networks in exploiting latest technologies or anomalies of PTT prices. Where, for example, long distance and international telephone calls continue to be charged at a very high premium there will be scope for efficient 'private use' and reselling of leased circuits.

KEEPING UP WITH TECHNOLOGY AND THE MARKET

You can only keep up with the capabilities of technology and have realistic expectation of market services and prices by being involved in the market. This requires active reviews from time to time. The best time for such reviews is always at a time of change—the review point of an existing contract, a business reorganisation, or a new venture. An individual review is most easily carried out in the context of what you need to do, and the information is then gathered by asking all possible prospective suppliers to tender a proposal.

The proposals might address a specific matter, such as the replacement of a particular, obsolete, piece of equipment (eg a company Private Branch telephone eXchange, or PBX). Alternatively, you could ask for proposals of a more general nature, such as 'for monitoring stock through a distribution function'. This might help as an input to the overall business planning process, stimulating new ideas. Increasingly the better suppliers are striving to offer solutions more of this nature. It is not important whether their ideas work in their initial form. What is important is to bring new stimulus and ideas to the business planning and management process, and enable these to germinate in a working group of mixed business folk and technologists.

Having built a good relationship with a relatively small number of suppliers, you will find that they start to understand your general business approach, adapt to it and proffer new ideas without waiting to be asked. You should encourage this. Such a joint approach with suppliers is probably the most productive way of achieving real edge on the competition through the use of telecommunication.

COMMON INTEREST GROUPS

Telecommunications managers associations and industry associations are another good way to keep up with the general market and latest news. For a telecom manager normally alone within his company, it is a good opportunity to meet with others in a similar position and facing common problems and opportunities. Recommendations and ideas are always forthcoming. There need be no risk of giving away confidential company secrets—you can still learn basic telecom management tricks and techniques.

Common interest groups can be a good way of applying pressure for change. Many large companies, for example, have used common interest groups in order to influence regulatory change—first pressing for it to happen, and then making input on what should be the result.

Exhibitions now and then, and specialised seminars, can help a telecom manager to meet others, see new technologies demonstrated, and be subjected to new ideas.

For a full time telecom manager, I believe it is important to spend at least some time amongst common interest groups and at exhibitions, but the amount of involvement needs to be balanced against short and long term benefits.

WHAT STANCE TO HAVE ON TECHNICAL STANDARDS

Is any standard actually standard? Most basic equipment is very well standardised. Nobody ever thinks twice about whether the telephone exchange will connect to the line. Few mistakes are possible at this level. But the most recently developed, so-called 'leading edge' equipment often does not conform to any agreed standards. Should anyone invest in such equipment, given the obvious risk of early obsolescence caused by incompatibility with other equipment? The simple answer is that you may have no choice. So let us explore how the situation arises, and how some protection can be achieved.

In some cases, the lack of an officially agreed standard is caused by insufficient elapsed time since the new discovery for the various aspects to be discussed openly in a public standards forum, documented, improved and

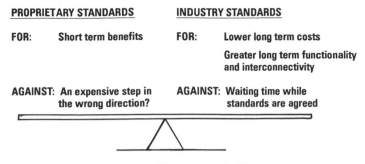

Figure 2.4   The standards dilemma

agreed. Manufacturers are understandably keen to try to recoup some of their development costs as early as possible, and thus may feel compelled to launch an interim product which pre-dates the standards. In these cases, some protection of your investment can be achieved by getting an assurance (perhaps with a monetary withholding) from the manufacturer that he will either upgrade your equipment once the standard is available, or replace it.

In other cases, companies may not wish to expose their technological developments to the public gaze, because of the intrinsic commercial value of their secret. The strategy may work for some companies, operating in niches of the market which they feel single-handedly able to dominate. (An example might be a line multiplier device—two identical units working at either end of a point-to-point link could use a non-standard interconnection method to provide increased line capacity. Several relatively small companies make devices of this type.) However, a manufacturer's isolationist strategy with respect to the technical standards of his products can also backfire if other companies are not able to develop complementary products. (Many 'proprietary' inter-PBX signalling systems are not much used, despite their huge potential, simply because it would require all of a company's PBXs to be converted to be of the same equipment type—at huge cost. The development money was thus largely wasted.)

Finally, there are those companies that do not submit their technical developments to become public standards because they do not need to. The size and standing of an inventing company is sometimes sufficient to establish their technical work as a standard. Historically, the best example of a company in this bracket has been IBM, though the current move towards open standards which allow different manufacturers' computer systems to intercommunicate is tending to undermine their previous stranglehold on some standards. Their SNA standard for data networking is published directly by themselves—not by any public body. If they do not choose to, there is no need to change their standard to accommodate others. So they could try to exploit the situation to reinforce sales of their own products.

So what's the advice?

(1) Try to stick to standards where you can. Where more than one 'standard' exists in a given field (eg the early videoconference encoding standards), secure by agreement with your business and technologist colleagues a 'company standard' which is the best compromise of value-for-money and maximum interconnectivity with outside suppliers and customers.

(2) Minimise investment prior to availability of the public standard, unless the short-term investment in its own right yields significant business advantage or competitive edge to justify itself.

(3) Always seek assurance from the supplier that he will upgrade or replace his equipment once the standard is agreed—and fix a ceiling price.

(4) But be careful not to be too rigid and inflexible with regard to standards used in the company. If you are, you will stifle the business in its innovation.

THE CYCLE OF IMPROVEMENT

We end the chapter with a few thoughts on telecom management technique. In particular, we present a few ideas on how to choose and structure the right projects to create a virtuous cycle of continuous improvement.

Consider the cycle of a project as illustrated in Figure 2.5. The ideal project delivers in all timescales—boxes 2, 3 and 4. In the short term (box 2), some sort of incentive (usually reduced cost) is necessary to persuade users that the project is worth supporting (that there is benefit in changing supplier, or whatever). If there are no benefits here, the project dies before it starts. But tactical benefits are likely to be soon eroded, so we need also to look for a longer term gain (box 3)—maybe extra business for the company through locked-in customers, or dramatically reduced costs through improved business efficiency.

Finally, the project needs to have a natural review point—a point in time when you will be better informed for a reassessment of the new options as they may have changed.

Only 'ideal' projects should be chosen. Or, perhaps more helpfully, projects should be structured in such a way that they fit the ideal model. Let us explain with the aid of an example.

In a previous job I worked for a major pub company, which had thousands of pubs scattered throughout the UK—many of which had a payphone. The problem was that the company had no idea how much

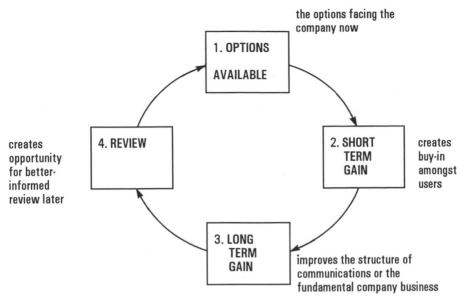

Figure 2.5   The ideal cycle of a well-structed project

money was made or lost through the payphones, or indeed exactly what payphones it had. The takings from the payphone were collected by the publican, and declared with his other weekly takings. But this did not help get a real picture of the pub payphones business. The problems were several layers deep:

(1) The company was not convinced all the takings were declared by individual publicans.

(2) There was no way of reconciling phone bills against takings, as the cut-off dates for takings declarations and for phonebills were nearly always on different days.

(3) Anyway the phone bill gave no clue about the 'unused' money which had been inserted in the box (the unused fraction of a 10p or £1 coin).

(4) Finally, losses through vandalism were also indeterminate. These would accrue both from direct losses and loss of business while telephones were out of order.

The challenge was to start a project which would get things back under control. The project chosen was a field trial of a 'managed payphone service'—offered by British Telecom (from whom we were already renting most of the boxes anyway).

Under the scheme, British Telecom would entirely take over the pay-phone service in a large cross-section of pubs—in effect providing an extension of their own public payphone service on our premises. They would be responsible for determining what equipment should be installed, and would carry the costs of installation, maintenance and coinbox emptying. In return, they would pay a percentage of the profits back to the pub.

In the short term, the pub company would dramatically improve their cashflow on the payphones—no more outlays at the beginning of the calendar quarter and revenues gradually accrued thereafter. Instead no outlays at all. Also, lost takings (publican theft, if any) could be recovered by the company.

In the medium term the public image of the payphone in the pub stood to be better—better looked after by a specialist telecom company rather than part-time looked after by a publican. There would be more incentive for British Telecom to be swift in their maintenance.

In the longer (review) timeframe, statistics of usage, takings, profits and costs (made available to us by British Telecom) would allow the pub company to be better informed when it faced further evolution of its payphone service.

What was in it for British Telecom? A cut of the increased profits, achieved through better management, and the retained profit of the under-lying telephone network business, which in consequence the pub company did not consider diverting to another supplier.

3
Human Communication Services — Voice and Video

Typically 90% of a company's expenditure on telecommunications is on 'plain old telephone service'. This is therefore an important area to understand and manage well, seeking maximum economy and effectiveness.

THE VALUE OF TELEPHONE CONVERSATION

One of the biggest difficulties facing the telecom manager of a company today is what value to associate with voice (ie telephone) communication. How much should therefore be spent? In future the problem will be the same for the videotelephone services as they replace straight telephone service.

Given that as much as 90% of the total company telecom bill can be for telephone services alone, for our imaginary $200 million turnover company this would be around $900k per annum. A lot of money to spend, but what can the company say it got in return? Unfortunately, I don't have a simple answer. Most companies could not survive without the telephone, but whether it is used too much or too little, and what proportion of the time (and expenditure) is used well may be a lot harder to assess.

Service-oriented companies should place a high value upon telephone services, for these provide a customer with immediacy of access to his supplier—and may be the only *human* contact with the company over long periods of time. A customer's experience of how customer-friendly the domestic gas company is may be based on a telephone call made two years ago. Two minutes of effort may represent two years of reputation!

If a customer has a question about available services or has a problem, then he needs to be dealt with competently and immediately. If he is in the mood for buying, use the mood. If he is in the mood for complaining, then his anger needs to be vented and a correcting action needs to be put in train. The telephone provides an instantaneous and human medium accepted by most customers.

If customers do not themselves call, why not give them a ring from time to time? By doing so you save on the cost of a personal visit, and get the chance for a friendly review of how well-suited your products are, and what opportunities exist for you to sell more of them. In addition, you may dramatically improve the customer's outlook on the customer-friendliness of your company, The call will be most beneficial if previously you have only had human contact with your customer when he has felt enraged sufficiently to motivate a complaint. Minor grumbles may have gone completely un-recognised, and sales opportunities will have been lost, if only due to the customer's unawareness of your full product capabilities.

For some telephone-oriented business functions, some quite definite performance measures can be developed. For example, a complaints or order-handling bureau can be judged on how quickly it works (ie the average duration of calls and the average time to answer). But for much business use, these measures have little meaning or value.

The challenge over time must be to improve the value achieved by general business calls, and to develop simple metrics which help to improve the effectiveness and efficiency of communication. Examples of questions that the telecom manager can address with departments are as follows:

With the personnel department:

• Can individuals become more effective through better communication?

• Can the organisation as a whole become more effective—closer-knit, and more team-spirited?

With the marketing department:

• Can customer service be radically improved through better contact and awareness of customers' needs?

• Can we become more available to our customers?

In his own department:

• What technologies help to encourage the right sort of communication, and to suppress bad practices?

• How many messages go to the wrong place? To a person on holiday rather than a department or function that could have handled them in his absence? To an individual's desk rather than to the individual himself (who may be somewhere else)? How many orders are handled by the 'wrong' department? Would the use of different technology help to improve any of these processes, make messages available to all members of a department, or enable messages to go straight to their intended recipients?

And so on...

The telecom manager's objective must be to achieve a clearer link between the expenditure on telephone communication and the benefit. Only in this way can we gain the awareness needed amongst users for its most effective and efficient use, and promote the use of alternative methods where they would be more appropriate.

THE COMPANY TELEPHONE SYSTEM

When companies have only a very small number of employees, individual telephone lines rented from the local telephone company are likely to meet all needs and be the best value for money. But with only a few more employees (and maybe when there are only two employees for some types of business), the company may have to start considering buying or renting a specialist telephone exchange system (such as a Private Branch Exchange, PBX, or 'key system'). Such systems enable the different employees to make better use of the external telephone lines, so that an individual external line per employee is not needed. They also enable incoming calls to be transferred easily between employees (Figure 3.1).

At the point in time when a small company needs its own on-site telephone exchange, it is best for it to obtain expert advice about which system to purchase and how much money to spend, since it is likely to be a substantial capital investment or rental cost.

For a small company, both advice and the PBX equipment is best obtained from the telephone company, for they are close at hand, and a low risk supplier. You can always threaten not to pay your telephone bill if they fail to service the equipment so you retain some control over them.

The decision as to whether you rent or buy a system should include consideration of capital against rental costs, service and maintenance needs, and a reflection upon the number of years for which you are likely to need the equipment. (Might your staff numbers quickly outgrow your equipment? Or might you soon want to upgrade to a more up-to-date version?) Uncertainty is best met by renting equipment.

A larger company will have more extensive needs—perhaps requiring to serve multiple sites with PBXs configured into a single network. The basic technology choices are the same as for a small company, but the larger

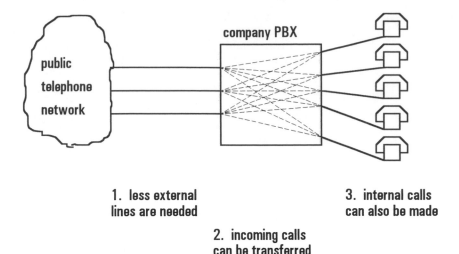

**1. less external
lines are needed**

**3. internal calls
can also be made**

**2. incoming calls
can be transferred**

Figure 3.1   The benefits of a company PBX telephone system

company may have a wider choice of supplier. When purchasing multiple PBXs, for example, a larger company can negotiate for lower prices, and use its size to demand the ongoing support which it may need from the supplier. For such companies, renting equipment from the telephone company may not be the cheapest or best solution. Instead a direct supply contract with an equipment manufacturer is likely to yield better price and service, for there is no middle man to be paid and no intermediate field technician to be trained.

CONSIDERATIONS WHEN INSTALLING A COMPANY TELEPHONE EXCHANGE

Today's PBX technology is both sophisticated and reliable. But while it may take specialist advice and training to design and install the system, it requires little further attention during normal operation.

The most important decisions for the company having a PBX installed must all be made prior to its installation. The main ones are listed in the table below:

CONSIDERATION	Remarks
Employee 'extension' lines	How many will be needed?
External lines	How many will be needed?

Special features	1. Is 'direct dialling-in' (DDI or DID) to employee extensions required, or will all incoming calls first be answered by the 'central operator'? The benefit of a central operator is your ability to ensure a human answer for all callers. The drawback is the additional staff needed. And beware, the benefit is lost if the operator is largely unhelpful—doesn't know where the employee is, isn't able to advise on who else might be able to help, proves unreliable in delivering messages, or worse still, simply refuses to take one. For frequently absent employees a voicemail system (discussed later in the chapter) may be a better alternative.
	2. Is barring of long distance, international and public telephone operator calls required?
	3. Are sophisticated user controls required, such as 'ring back when free', conference calls, etc?
	4. If more than one site is being equipped, will there be a single 'central operator' for all sites, or will features like 'ring back when free' be required between sites? If so, some sophisticated inter-PBX signalling may be necessary, and a 'tie-line' network may need to be included as part of the system design.
	5. Will a call logging system be required (to monitor dialled destinations and call durations and costs)?
	6. Is particularly heavy incoming calling expected to any individual person or department? If so, a call queueing system such as ACD (automatic call distribution) equipment may be worth considering.
Maintenance and support	1. Will one of your own staff perform the basic administration (ie allocating telephone numbers, and reallocating these to different offices when staff move around but wish to keep the same number). Or will the supplier do this?
	2. Who will diagnose and fix the equipment when it goes wrong, and how long will it take them? Are they just a short travelling distance away?
Buy or Rent?	Renting is a better insurance against possible changes, but purchase is likely to be cheaper in the long run. When calculating comparable costs, remember that rental charges usually include ongoing maintenance charges, so include a similar figure in your 'purchase option'.
Maximum foreseeable needs	Many systems are modular, and can relatively easily be expanded to accommodate your future needs. However, all systems have a maximum capacity, and to exceed this would require replacement of the system with different equipment. Avoid this situation by buying an initial system which is capable of expansion beyond all foreseeable needs.

MONITORING TELEPHONE NETWORK USAGE

Monitoring telephone usage is important for three main reasons:

(1) Determining current demand—and associated costs and unit costs;

(2) Ensuring adequate quality of service (calls getting through and being answered);

(3) Confirming that the technology and network being used are those that will give minimum cost.

The pattern of general telephone usage can be a good indicator of how well a company is operating—revealing how frequently the various departments inter-communicate or talk with customers and suppliers. To know what is actually being said may not be important. High usage may reveal good accord between departments, or belie the failure of some other communications channel. (A lot of calls from distribution to sales to clarify incorrect shipping advices might signal either incorrect availability of information to sales, incorrect typing into the computer, or some other problem.)

High usage may present an opportunity to convert economically from telephone as the prime or only communication means for a particular purpose to a newer and better suited electronic or data technology (eg automated ordering). Conversely, very low usage between departments might reveal lack of cooperation or conflict, and might also signal a need for intervention.

HOW TO MONITOR TELEPHONE NETWORK USAGE

There are two main sources for information about telephone usage. These are:

(1) the telephone bill;

(2) a call-logging system connected to the company PBX.

The telephone bill (Figure 3.2) is obviously the best method of determining overall cost. But unfortunately, not all telephone companies are in the habit of providing much detail on the breakdown of costs. The more advanced telephone companies now offer fully itemised call records (Figure 3.3)—giving the time, destination and duration of all calls made—but the laggards give only a breakdown of equipment rental and then a total figure for calls during the month or calendar quarter. Frankly, I have not found either format ideal. The second case gives far too little information, but the fully itemised bill can be too much if it's not in exactly the right internal cost centre structure (apart from the sensitivity of senior management to the telecom manager knowing where their calls were going to, consider trying to process a paper record of all the telephone calls of a large corporation).

Moves are now afoot in several telephone companies to provide more suitable information on computer floppy disk. Simultaneous availability of a PC with suitable software to analyse the calling pattern provided on floppy disk will represent a major step forward.

From the company's PBX, computer-readable information is probably already available—for sites that have call loggers. But given the cost of some call loggers ($10k upwards), a single portable device may have to suffice, alternated between different sites to give a sample of calls. Knowing about every single call is not important.

XYZ Telephone				
Company plc		Date of bill	Charge Period	
		31.07.93	July 1993	

Article Number	*Article*	*Quantity*	*Unit Charge*	*Total Charge*
10130	Telephone line rental, July	1	24.60	24.60
17120	Telephone call units	750	0.23	172.50
91205	Telephone handset rental	1	3.07	3.07
			Total this month ex VAT	200.17
	09 21415237-1234	07.93	VAT free	197.10
	1234000012234	1,00	Liable to VAT	3.07
			VAT payable 15 %	0.46
	Martin Clark		Total this month (incl VAT)	200.63
	1 The Mews			
	London		Outstanding from previous period	0.00
	You can call us at:		**Total to Pay**	200.63
	071 234 5678			

Figure 3.2   A typical unitemised telephone company bill

Detail of dialled calls				
Date	Time	Destination	Duration HR:MN:SC	Amount
01/07	08:05	0452 860000	00:12:56	0.63
01/07	08:20	0103314559900	00:02:08	1.80
01/07	08:27	0481676767	00:34:30	2.18
01/07	09:20	010496543210	00:04:49	1.39
01/07	09:40	0734654321	00:29:31	1.42
01/07	10:25	0819876543	00:13:15	0.42
01/07	10:47	0718765432	00:13:53	0.58
01/07	11:12	0606765432	00:09:45	0.50
01/07	11:31	0202456789	00:08.19	0.42
01/07	11:53	0432765432	00:06:01	0.42
01/07	12:06	0635754321	00:07.55	0.84
01/07	12:18	0954987654	00:11:51	0.58
01/07	12:34	01012123456789	00:29:34	4.78
01/07	13:23	010616548764	00:24:23	5.67
01/07	14:06	0734123456	00:29:23	1.83
...and so on...				

Figure 3.3   Sample of a typical telephone company call itemisation

WHAT TO LOOK FOR AND WHY

What we are trying to find out is who, in the main, the company is talking to, and why, and how much it is costing, in order to assess the potential for using better suited technology or better value suppliers.

The call statistics should first be sorted into totals of calls to the major destinations. In particular we need to check for a large number of calls to the same site (say a company factory, a particular customer or a supplier) or to a particular geographic area (a particular town, region or overseas country). This is done by comparing the first few digits of the dialled number. (Indeed some call logs only give the first few digits as a way of protecting the confidentiality of individual callers.)

What we seek is to determine those destinations with particularly heavy traffic—ie those for which special network arrangements or other technologies may be justified. In particular we need to find out the total number of calls made and the associated cost, as well as the pattern of calling during the day (high and low usage periods as shown in Figure 3.4).

It is the pattern of calling, and in particular the peak calling rate (measured in Erlangs, the number of simultaneous calls) which is the greatest determinator of cost when building a private network. Buying the circuit is what costs the money; you can use it as much as you like thereafter without further expense. By contrast, on a public network, the total number of calls made is the cost determinator. Thus where a large number of calls are made each day, but within a relatively small time window, carriage over the public telephone network may be the cheapest option. But where a relatively low intensity of calling is maintained all day, a direct, private network connection may be warranted.

A truncated peak appearing in the traffic pattern may be a signal of congestion—too few circuits to cope with the demand, too few people to answer the calls at the other end, or simply too few lines provided at the distant site. This can happen on tie line connections between sites, but can also occur on the main connections from the company PBX to the public telephone network. When it does happen, there will be considerable management frustration, not to mention the impatience of customers trying to call in. People nowadays don't expect to get no line or continuous busy when they try to use the telephone.

In some circumstances it may also be wise to check some of the answering

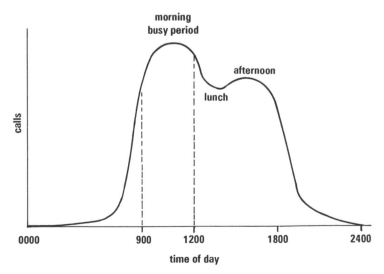

Figure 3.4   Typical office calling pattern

rates on incoming calls. People calling the main telephone reception number, or a service bureau, expect to be promptly answered. Useful statistics to measure are:

(1) the overall percentage answer rate;

(2) the average time to answer;

(3) the percentage of calls answered within a given target time (say 10 seconds).

Where the time to answer is prone to prolongation, an arrangement should be made for some kind of automatic answering, like those used by the telephone companies in their directory enquiry service bureau ('you have been connected to the directory enquiry service, please hold the line until an agent is available'). This confirms to callers that they have called the correct number and that a further short wait is worthwhile.

A LITTLE ABOUT TELEPHONE TRAFFIC STATISTICS

In a telephone network a connection is made between caller and destination by the establishment of an end-to-end chain of exchanges and interconnecting circuits. During the call set-up, each exchange in turn determines where the call is destined for and selects the best available circuit from it to the next exchange, provided one is free. It is normal to provide sufficient circuits within the network so that the probability of unavailability of a line to the destination is about 1%. Any rate higher than this is likely to cause frustration. There is, of course, a much higher probability that the person being called is already on another call, but callers accept this. What is less acceptable is the announcement 'all lines from Oxford to London are engaged'.

To maintain an average of five simultaneous calls (in shorthand, 5 Erlangs) between two sites during the busiest hour of the day, about ten circuits are required if the grade of service (ie the probability of lost calls) is to be kept below 1%. The extra circuits are needed because although the average is five, fluctuation in call-in rate will mean that typically between three and nine circuits are busy at any one moment. A whole science goes with this subject and, for those interested readers, might I refer you to my earlier book *Networks and Telecommunications* (Wiley, 1991).

Suffice it here to say that the number of circuits required to meet the traffic demand between adjacent exchanges in a private network, and the number between a PBX and the public exchange, needs to be re-calculated on a regular basis. Circuit numbers need to exceed average peak hour traffic (ie the average number of simultaneous calls—measured in Erlangs) by around 5 circuits if unacceptable congestion is not to result.

RESTRAINING TELEPHONE COSTS

An element of straightforward cost restraint is necessary in order to keep telephone usage costs, in particular, under control and free of wild fluctuation, but you need to be careful that this does not become a sole obsession. The restraints I recommend are those which discourage obvious misuse. Most PBXs offer the opportunity to bar calls to trunk and/or international destinations, to the public operator and to high cost information services (such as the sports score line). These are obvious precautions. Personally however, I do not favour the sort of regular 'witch hunts' which seek to catch individuals making personal phone calls. To me this is too damaging to general staff morale—staff who don't feel trusted react with lower pride and responsibility in their work.

PRIVATE AND PUBLIC NETWORK OPTIONS AND THEIR ECONOMIC DIFFERENCES

For large companies with significant internal company telephone traffic spanning more than one site, there are two options for the carriage of telephone calls between company sites:

(1) the public telephone network, or

(2) a so-called *private network*, comprising company PBXs interlinked over direct circuit connections leased from the telephone company. (Actually installing transmission lines between your sites, even if by radio, is forbidden by law in most countries.)

In effect, for any company that has a private network or is considering one, there is an eternal question requiring answer—'Is a private or a public telephone network most economic and beneficial, given today's capabilities and prices?' We discuss here the main considerations.

All companies will need to use the public telephone network to some extent, if only to contact and be contacted by multifarious external parties. The only question, therefore, is whether there is an economic or other benefit from using a private network to handle some or all of the internal company traffic.

First, let us look at the possible justifications in the case of point-to-point traffic between two sites, using the example illustrated in Figure 3.5.

The financial comparison for point-to-point traffic is relatively straightforward.

- The total minutes T multiplied by the relevant public tariff rate, gives the cost of using the public network.

- The comparative private network cost is obtained by calculating first the required number of circuits from the peak traffic E (in practice around $E + 5$). This number is then multiplied by the annual leasing charge for a circuit between the two sites.

Having justified the link on economic grounds, further benefits may accrue from the use of a sophisticated inter-PBX signalling system between the two sites. This may, for example, open the opportunities for:

- A common operator to be used for handling incoming calls to both sites.

- Common numbering scheme, with short digit dialling between extensions.

- Other features such as 'hold on busy', 'ring back when free', 'conferencing'.

Alternatively a hybrid public/private network solution may offer overall best value for money. In this solution, too few direct private wires are installed, and the remainder of peak hour traffic is 'overflowed' via the public network as shown in Figure 3.6. There is a slight drawback. You will not be able to offer the 'ring back when free' facility reliably any more, and the common numbering scheme of the private network solution may suffer. You'll have to judge for yourself whether this is a real problem or not. But for one, I've never convinced myself of the need to spend a premium in order to get linked numbering.

When the network covers several sites, the complexity of the potential solutions increases quickly. Consider five sites and a private network configured between them in a star topology as illustrated in Figure 3.7. Now the pattern of calls to be studied is actually quite a complicated table of to and

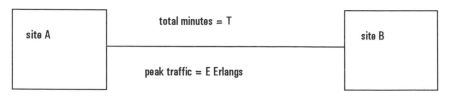

Figure 3.5   Telephone traffic between two sites

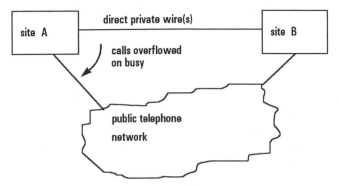

Figure 3.6   Hybrid private/public network link

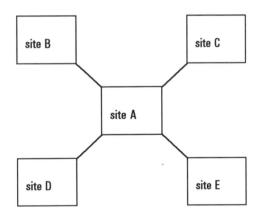

Figure 3.7   Possible private network configuration comprising five nodes

from traffic streams between each node and the four others. The public network cost is still easy to calculate. But the private network cost calculation is more difficult. For example, traffic from site D to site E now passes over two links, DA and AE. It might of course be cheaper to install a direct link DE, but this link could not be used by traffic from either site B or site C. But there again, maybe site A is not the most economic centre of the star!

At the start, it may be possible to design an optimum configuration, but as time progresses, the optimum may become hard to maintain, as new circuits are added incrementally to an existing infrastructure to accommodate changes in the overall traffic pattern. As the network gets more complicated, a call across the private network might actually traverse four or five separate links—with both a cost and potential quality impact.

VPN, OR VARIATIONS ON A PUBLIC NETWORK THEME

A recent variation on the theme of public telephone network service is that of VPN, or in its full name Virtual Private Network. This began life as a technological concept. The theory was that a large corporation could achieve the rationalised numbering scheme benefits, and perhaps some of the 'ring back when free' type features without actually building a private network. Instead the effect of a private network would be created within the public telephone network—literally a 'Virtual' private network—but without the commitment of customer capital.

While these service type benefits may be of value in their own right, what has more recently greatly increased the interest in virtual private networks has been the massive discounts on public telephone tariffs made available by some of the telephone companies. Typical in the United States over the last few years, have been VPN deals with major customers being given up to 60% discount on normal published tariffs, where the customer has significant overall volume.

A FEW THINGS TO TRY IF YOU HAVE A PRIVATE TELEPHONE NETWORK

Private networks will tend to be most cost effective in countries with relatively high telephone tariffs, but disproportionately low leased-line charges. In general, this sort of discrepancy only exists in the countries where tariffs are not being forced closer to cost either by competition or government action. These are the countries in which to explore private network development (if it is legal).

In countries where the telecom market is being advanced rapidly by competition (eg the United States), telephone company margins on long distance calls have already been significantly reduced, so that pricing overall is much more cost-oriented. In these countries, the scope for economic private networking is much reduced, but may still exist if there is particularly heavy traffic demand.

Where you do have particularly heavy traffic between specific sites or regions within the private network, there may be further scope for cost reduction, by the use of voice compression equipment—using specialised equipment literally to multiply the capacity of your leased lines by a factor between two and five. So for each real circuit, between two and five calls can be carried. Note, however, that the quality may suffer, particularly if an individual call is compressed and expanded several times on its passage through the network. The most obvious applications of voice compression are on international leaselines (eg transatlantic ones) or long distance national ones (say, coast to coast across the United States). On very short lines, the cost of the compression equipment may be more than the cost of buying the extra lines.

THINGS TO AVOID—THE PITFALLS OF PRIVATE TELEPHONE NETWORKS

Once private networks get relatively large, they become increasingly complicated to manage. Add-on facilities to the network, added and justified at marginal cost, can start to become the *raison d'être* for the network as a whole. Continuously optimised costs are difficult to achieve on networks with changing demand, and the base of existing assets may restrict flexibility for change.

Another major trap which has ensnared some companies is the difficulty and cost of reselling telephone network service using the private network to carry the calls for third parties (in competition with the public telephone network). This should only be contemplated on immense networks or in the case of very limited customer calling needs. The problem is that the callers and the people they want to call are bound not to be directly connected to your network. The total cost to them will therefore include (in one form or another) the cost of two public telephone network calls, as well as your network costs. One call is made by the call originator across the public network to reach your network. The second public network call your network has to make is in order to deliver the call to its final destination. Two public network calls and the associated costs—but only one call made as far as your customer is concerned. Be careful also about the economics of any public network bypass that you attempt for your own traffic (illustrated in Figure 3.8).

Still more danger awaits those private network designers who attempt so-called Off-Net to Off-Net service (otherwise known as call-in/call-out). The theory of it goes like this. The cost of international (or long distance) calls across the private network or VPN between company sites (eg in the UK and USA) is much cheaper than the public network tariff for the same calls. So why not capitalise on this fact for calls between sites in the USA and UK even though such sites may not be connected to the private network? This would work by first making a relatively cheap 'local' call to connect to the private network in the United States. The call would then 'ride' the private network to the UK, where another 'local' call made across the public network would see it through to its final destination.

Well, the theory may work in practice as well, but you need first to confirm:

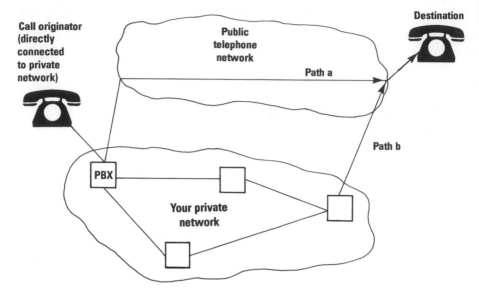

Figure 3.8   The principle of public network bypass

(1) That such action is legal;

(2) That the cost of the two national calls (in the UK and USA), added to the private network costs, makes economic sense;

(3) That the quality of the connection will be adequate;

(4) That you have precautions in place to protect you against fraud, since potentially everyone in the USA has the ability to call anyone in the UK over your network! Only a couple of days of these costs could be quite ruinous.

ADVANCED VOICE NETWORK SERVICES AND THEIR BUSINESS CONTEXT

Apart from determining how best to meet their basic telephone network needs, companies today are faced with a number of decisions about which other voice-based communications services and value-added services are relevant to their business. We shall discuss each of the following:

- Mobile telephony (cellular radio)
- Cordless telephony, and the evolution towards personal networks

- Freephone (reverse charging) and information services
- Calling line identity (identifying callers before you answer)
- Voicemail
- Voice response systems
- Calling card service
- Tele-conferencing.

Mobile Telephony

Cellular radio (mobile and car telephones) has been available from many national telephone companies now since the early 1980s. It allows users to make and receive telephone calls from most geographical areas within a given country. It has grown rapidly although its relatively high cost of usage (typically about three times as expensive as ordinary telephone calls) has tended to restrict use of mobile and carphones to certain specific user groups:

- Senior company managers;

- Field sales staff and sales management;

- Self-employed individuals who work mainly on client premises (eg local building or maintenance contractors);

- Lorry (truck) distribution fleets, for despatch monitoring.

In the early days there were problems with the quality of the service. There were high congestion levels when trying to make new calls, and sometimes calls were cut off in mid conversation. These problems were largely due to the fact that initial demand for the phones outstripped the rate of infrastructure development, and certain geographical areas were very poorly covered with radio base stations. Most of these problems are now behind us, but a new problem is looming. The first generation of *analogue* networks and handsets are inefficient in their use of radio bandwidth (a limited natural resource), and the maximum capacity for call making may soon be reached in many areas. The networks will therefore progressively reach their saturation numbers.

Another limitation, particularly felt in larger landmasses (like continental Europe), is the technical incompatibility of different national cellular radio networks. This means that while a German with a 'C-Net' phone in his car might think nothing of driving to his business meetings in Brussels, Amsterdam or Strasbourg, his phone would not work when he got there—no chance to call the office, and they can't contact him either.

Because of these reasons, a new generation of cellular radio is emerging—one based on more recent *digital* technology. The new system, called GSM (Global System for Mobilecommunication), will be much more efficient in its use of radio bandwidth, will give higher quality of speech transmission, and will allow 'roaming' between different operators' networks. Thus, for example, a German user on the GSM system (he would be a subscriber of either 'D1' or 'D2' in Germany) will ultimately be able to drive anywhere in the European Community and still be able to use his phone.

Once mobile telephony begins to reach this stage of development, I believe companies should seriously consider its more widespread use, for mobile telephones are much more effective at ensuring contact with the individual rather than the mere transmission of a message to his desk.

Initially, GSM seems likely to be priced about the same as existing analogue cellular radio—since there is still latent demand, but in the longer term some operators are proclaiming significant reductions in price as the result of larger scale production.

Some experts view the cellular technology (whether analogue or digital) as inherently too costly to create a mass market. For these soothsayers, the mass market which will put a phone in everyone's pocket will be the 'cordless telephony' market, which we discuss next. Either way, it seems likely that one or both of these technologies will greatly expand business use of mobile telephony.

Cordless Telephony, and the Evolution Towards Personal Networks

To many people today, cordless telephony (if it means anything) may conjure up the image of a home telephone, with which users are able to walk around the house and garden, or perhaps the office, while making calls. In fact, used just like this, cordless telephones are becoming very popular—cordless telephones already represent more than 55% of telephone sales in Japan, and other markets are quickly expected to follow suit.

But cordless telephony can offer much more, and second and third generation cordless telephone technologies are already in evolution. These aim to be the 'mobile phone of the masses'—only incrementally more expensive than today's 'fixed' telephones, but far more flexible. The goal is a personal phone, with a personal number, in everyone's pocket—and public antennas all over the place—to enable you to use it anywhere. You'll be able to send and receive calls from anywhere in the land, but may have to stand still or move only slowly while in conversation—not quite as 'mobile' as today's car telephones, but aiming to be far cheaper.

Should these technologies take off, then we can expect everyone in

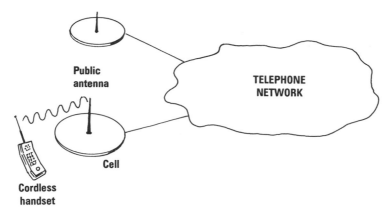

Figure 3.9   A public cordless telephone service

business to have one. The office phone system and its constraints will disappear. No more calling someone's desk just to find he's not at it, you'll be able to call him directly. A network between people rather than between places.

Early forms of the new technology have appeared in several countries — CT2 or Telepoint in the UK, Pointel in France, PCN (Personal Communications Network) and DECT (Digital European Cordless Telephony) in the European Community, Personal Handi-Phone (PHP) in Japan. In addition, wireless office telephone systems to ease the problem of the 'man stepped away from his desk'. The diversity of technology currently available is a signal of the industrial interest in the market, but is probably also a warning that there is yet some way to go in the technological development.

Some ventures have already tried and failed (for example, some Telepoint operators in the UK). But the failure of the extinct operators to some extent came from an unduly hasty rush to offer service before their products were properly ready. The handsets turned out to be too expensive and the batteries in them to be very short-lived. The service was somewhat unreliable, geographically patchy and not as cheap as it was claimed it would be. Despite this failure, I am confident that something much more worthwhile will come of all the effort within the next ten years.

Freephone (Reverse Charging) and Information Services

Automatic freephone is a service that enables a company to pay the call charges incurred by their incoming callers. The service is attractive to companies in a number of instances:

- Encouraging prospective customers to call, perhaps in response to an advertisement, by removing the 'barrier' of the call charge.

- Encouraging existing customers to call helpdesks to report problems, rather than let the problems grow and result in lost business.

- Enabling field operations staff or travelling managers to call back into the office without the normal inconvenience of finding change to feed the public payphone, and the headache of recharging the expenses.

The Burger King company, for example, has a very successful 'Careline' service, which customers may call to express their opinions (good or bad) about their experience of Burger King restaurant service. When good opinions are consistently expressed, a particular restaurant manager may get some sort of reward. When problems arise, these can be quickly dealt with. The Careline service has also provided a very useful opportunity for well-targetted market research—the Careline operator is able to end the call by asking a small number of market research questions to people whom they know to be customers.

At the other extreme from freephone services, telephone companies also offer 'information service' numbers. When one of these numbers is called, the caller pays a premium charge over and above the normal call charge. The premium is, in effect, the charge which the caller is paying for the information he receives during the call. The premium is passed on by the telephone company to the company receiving the call. Examples of information services of this type are the 0898 service in the UK and the 900 service in the United States. Examples of the services that have been made available to callers are:

- accurate time
- weather reports
- sports results
- financial markets reports
- dating service, etc.

Clearly these services are not of interest to all companies, but may provide a valuable outlet for sale of information, particularly in cases where the information only has a short lifetime.

When using either freephone or information service numbers, companies must recognise the need to make provision for adequate simultaneous call handling, and for strict queueing of incoming calls. The queueing itself is usually performed using a (relatively expensive) piece of Automatic Call Distribution (ACD) equipment. This can cost upwards of $30k, although software to convert company PBXs to ACDs has recently started to bring this cost down. The advantage of an ACD is not only the ability to handle simultaneous call queueing and answering in turn, but also its capability to record relevant statistics about overall incoming call volumes and answer rates. It is an invaluable tool for telephone bureau management.

Before installing any kind of incoming call bureau, however, a company should have predicted its expected incoming call volume and know its target answer rate (say 90% of calls answered within 10 seconds). It can then calculate the required number of positions and operators. (More advice on how to do this appears in my earlier book *Networks and Telecommunications* (Wiley, 1991).)

Calling Line Identity (to Identify Callers Before Answer)

Some telephone centres answering incoming calls are already using a service called Calling Line Identity (CLI). This is a feature of the most advanced telephone networks (called *Integrated Services Digital Networks* or ISDNs), which enables call receivers to identify the telephone number of the caller prior to answer.

The availability of CLI could help the receiver decide whether to answer the call or not. But perhaps a more commercially minded application would be:

- The automatic display of a calling customer's account details and status on the receiving operator's computer terminal. This might allow the operator even to answer 'Good morning, Mr Smith, what can I do for you' — and so save the first few seconds of the call.

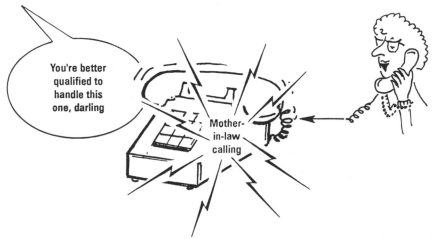

Figure 3.10   Using calling line identity to select the right customer service agent

● Alternatively, the device could be used to trace the identity of potential sales leads, from callers answering an advert.

There will be questions of 'data' and 'personal privacy' to be resolved by any company contemplating the use of CLI service.

Voicemail

Voicemail is perhaps most easily explained as a sort of 'second generation answerphone'—but is a much more powerful service. It is available in two main variant forms:

● public bureau voicemail service, or

● private voicemail service—usually linked to a company PBX or private telephone network.

Public bureau voicemail service was initially most common. Private network service is now beginning to take over the greater market share.

Voicemail suits users within a business community of interest, who might otherwise find it difficult to communicate, either because of:

● being permanently on the move (eg field service or sales staff)

or because of:

● international time-zone separation (there is only a very short and inconvenient time window for somebody in the UK to telephone someone in Australia).

The service works as if it were a central bank of answerphones. A caller wishing to leave a message calls the central 'voicebank' and leaves a verbal message for his addressee. This is usually quite short and informal in style. Subsequently, the addressee, who should be in the habit of calling the voicebank several times during the day, may find a number of short messages from various business colleagues on doing so. The system replays each message in turn and, immediately following each one, offers a number of options:

(1) To reply to the caller, perhaps copying the message to some other interested individuals.

(2) To forward the message, perhaps to a subordinate for some action.

(3) To store the message for later consideration.

(4) To note mentally the message and then delete it.

To execute any of these actions requires only a few simple key actions on either a tone-type telephone or on a special pocket-sized tone keypad. The user is saved the inconvenience of ringing back any of his callers—and manages to handle all of their enquiries at a time suitable to himself, rather than being disturbed by the telephone in the middle of an important meeting.

Keen users of voicemail point to the ease with which broadcast messages may be generated—perhaps the weekly report to the troops. In addition, they praise the direct and informal nature of the messages which users soon get in the habit of leaving. Unlike the telephone, there is no need for chit-chat and social niceties—just straight down to the business. So time is also saved.

An example of the effective use of voicemail might be its application in a sales force. The regional sales manager is easily able to broadcast targets and this week's news. The individual salesmen can report back orders, or perhaps refer customer questions.

More recently, it has become common for large companies to tie a voicemailbox to each employee's telephone line extension. This gives the mailbox an added dimension. Now mailbox owners can have their incoming telephone messages diverted to the mailbox, either when away from their desk or when busy on other important business and cannot be disturbed. In replying to messages, they can either call the person directly or may send a message to that person's voicemailbox. And when out of the office, they may call to a special telephone number to deal with the messages in their box. Provided users check their mailbox frequently (several times per day), even the most-travelled employee may appear to be only a few yards from his or her desk.

Voicemail, however, can fall into disrepute. The problem lies in the fact that most people take some getting used to leaving answerphone-type messages. Initially callers tend to hang up without leaving a message, and when they do it may simply be a request to call back. Should a caller feel his addressee is never at his desk, or worse, is at his desk but can never be bothered to answer the phone personally, then he can become extremely frustrated. I have seen senior management edicts banning the use of voicemail for this reason.

Voice Response Systems

A further potential of voicemail technology is in providing voice-prompted control of systems. On calling a given telephone number, for example, a voice response system could answer and ask the caller the nature of his call:

'Do you want our customer service department (dial 1), our sales depart-ment (dial 2), an account enquiry (dial 3) or some other matter (dial 0 for human assistance)?' According to the caller's dialled response, the system could either:

- Provide answers itself to specific caller enquiries.

- Ask further questions.

- Direct the call to an appropriate department, or specific individual.

- Record details (eg of a customer order)—either 'dialled' by the customer and stored directly in a computer, or stored in a verbal form for handling by a human. This might allow 24-hour and weekend ordering, for example.

In North America voice response voicemail systems are already very com-mon. You can, for example, call Montreal airport, punch in the flight number and get up-to-date news on flight arrival or departure times. Meanwhile, Air Canada's frequent flyer programme allows you in your next call to check your latest mileage credit, find out about award claims and this month's special offers, or merely request further literature.

Service information bureaux allow you to listen to the latest weather forecast, sports reports, snow depths and even the recommended wax to be used on cross-country ski trails.

Applications in Europe are currently only in their early stages of develop-ment, but are evolving fast. Initial European applications have included the delivery confirmation and fleet management of trucking distribution companies, simple order process systems, and 'noticeboard' systems for police 'community neighbourhood watch' schemes. Voice response systems in Europe continue to be held back by the relatively low penetration of tone-signalling phones—though hand-held signalling units have provided an adequate alternative for some of the applications.

Calling Card Service

AT&T in the United States have been international leaders in the introduc-tion of telephone calling cards. The cards make the business of making telephone calls much easier for people out of the office or away from home—either out for the day in their car, staying away overnight in a hotel or travelling internationally. The cards offer:

- Cashless calls made via the operator (or automatic equivalent) to almost anywhere, from almost any phone.

- Cost of calls itemised and charged directly back to the company, so avoiding the need to claim expenses.

- Call tariffs far below the heavily-marked-up prices often charged in hotels.

Teleconferencing

Teleconferencing is the name given to telephone calls involving participants in more than two locations. A meeting-like discussion can thus be held relatively cheaply (ie without travelling) and at relatively short notice.

The service is available on some company telephone exchanges, or otherwise from the national telephone company.

Like voicemail, it's a technique that takes a bit of getting used to. It differs from a meeting in that you cannot see any of the other participants, and this tends to have two disadvantageous effects:

(1) You can't tell who is talking or who else wants to express an opinion, so to avoid shouting over one another you end up with a moderator asking individuals in turn for their opinion (of course, this may be a benefit in giving everyone a more equal opportunity to talk).

(2) Some people can be a bit intimidated by the idea of an 'audience' whose reaction they cannot see.

Videoconferencing and Picture Telephony

Videoconferencing was the name given to the first real type of visual telephony. During a *videoconference*, two or more videoconference meeting rooms are connected together by high speed telecommunications links. A participant in one of the rooms sits at a table facing usually two or more television screens which are displaying the distant meeting participants in the various other locations. Alongside the TV screens, video cameras and microphones are recording the picture and sound that the distant participants will see and hear.

Picture and sound quality, while quite high, are not as good as normal television. Full television would simply be too expensive. International videoconferencing even in this form can cost as much as $2000 per hour!

Videoconferencing comes a lot closer than teleconferencing to recreating a meeting atmosphere despite what might be enormous geographic separation. Special camera equipment can be used to transmit shots of a white-board, of an overhead projector or 35mm slide illustration, or even of a video film. So the interaction can really be intensive and wide-ranging. However, in common with teleconferencing, videoconferencing is still an acquired skill, although groups who meet often using this means soon forget that it is not a face-to-face meeting.

Today's constraint on the use of videoconferencing has been the extremely high cost. A specialised studio meeting room can cost upwards of $100k to install, and the running costs of the communications lines can be as high as $2000 per hour. But even despite these costs, the medium has found favour in the biggest multinational companies as a means of saving on the stress and cost of travel for its most senior executives, and also as a way of promoting greater interaction between more junior employees in widely separated locations, who might otherwise not meet. The Digital Equipment Company (DEC), for example, uses videoconferencing to allow its software design engineers across the world to develop concepts and products far more quickly.

The future of videoconferencing is in miniaturisation and cost reduction. The new communications standard for videoconferencing (CCITT H261) will mean that devices made by different manufacturers can intercommunicate (whereas this was previously not possible). It will also make possible relatively high quality picture and sound transmission on standard digital ISDN telephone lines. This will greatly boost the usage of the devices and much reduce the usage costs, and so, in turn, is bound to spur greater efforts by videotelephone manufacturers to bring the costs down and increase the market size.

4

Networks for Computers and Electronic Equipment

Given that company business is becoming totally dependent on computer processing and associated data networks, far more general business managers should have a better appreciation of data communications and the likely problems. In the early part of this chapter, I have therefore elected to try to cover the technology in a little more depth than I have done for technologies in other chapters, explaining the various terms and techniques a little more fully, so that the subject-phobia may be reduced. The second part is more about practical management of data networks.

A LITTLE ABOUT HOW COMPUTERS 'TALK'

'Data' is the term used to describe information which is stored in and processed by computers. Data is stored in a form equivalent to a large number of electrical 'off' and 'on' states, and in this way either numbers or alphanumeric characters may be represented. Communication between computers can then be effected first by constructing an electrical line, and then transmitting the characters by some sort of rapid on and off switching mechanism.

Thus for example, in one of the common computer language codes called ASCII (American Standard Code for Information Interchange), we can represent the digits 0–9, and the letters A–Z, a–z, in the following form:

DIGIT OR ALPHANUMERIC	ASCII code	Transmitted as
	(so-called 'most significant bit' written first, as is normal convention—0 is equivalent to 'off', '1' to 'on')	(By convention, the 'least significant bit' is always transmitted first, so the on/off train is in the reverse order to the previous column)
digit 0	0110000	off-off-off-off-on-on-off
1	0110001	on-off-off-off-on-on-off
2	0110010	off-on-off-off-on-on-off
3	0110011	on-on-off-off-on-on-off
4	0110100	off-off-on-off-on-on-off
5	0110101	on-off-on-off-on-on-off
9	0111001	on-off-off-on-on-on-off
letter A	1000001	on-off-off-off-off-off-on
B	1000010	off-on-off-off-off-off-on
C	1000011	on-on-off-off-off-off-on
D	1000100	off-off-on-off-off-off-on
Z	1011010	off-on-off-on-on-off-on
a	1100001	on-off-off-off-off-on-on
b	1100010	off-on-off-off-off-on-on
c	1100011	on-on-off-off-off-on-on
d	1100100	off-off-on-off-off-on-on
z	1111010	off-on-off-on-on-on-on
etc	etc	etc

You will see from the above table how almost anything (including pictures made up of different colour dots (pixels)) could be represented in this on/off or 1/0 'data' style. In the code illustrated the standard is to use seven on/off (1 or 0) bits, but other computer language codes use eight or even more bits. Another commonly used code, EBCDIC (Extended Binary Coded Decimal Interchange Code), for example, uses eight bits per character.

All the on and off states, of course, have to be of equal time duration, so that the receiver can determine how many consecutive ons or offs have been sent. The number of on and off states sent during 1 second is called the bitrate, linespeed or transmission rate of the data communications line. Typical rates used in data networks are 2400 bits/second (written bit/s), 9600 bit/s (also written 9.6 kbit/s), 64 kbit/s (64 thousand bits per second), 1.5 Mbit/s (1.5 million bits per second), and 2 Mbit/s. The faster the rate, the faster the information can be transmitted.

But knowing the transmission rate of the line, and knowing which computer code (eg ASCII or EBCDIC) is being used, will not be quite enough information for a computer to properly de-scramble a message received in this manner. What is missing is the protocol—the etiquette of computer conversation—ensuring each computer talks only when the other is ready to receive, and providing a means for delimiting the characters and checking their correct receipt. If it were not for the protocol, for example, the following bit train could not be correctly interpreted:

0 1 1 0 0 0 1 1 1 0 0 1 0 0 → this bit transmitted first

You might think the interpretation is easy: 1100100 (the right-hand seven digits) is the letter 'd', while 0110001 is the digit '1', so that it could indeed be 'd1'. But we also know that 0111001 is the digit '9', so another possible interpretation is a character sequence 'something–9–something', where we don't currently have the beginning of the bit sequence for the first character or the end of the bit sequence for the last character. There are other interpretations.

THE IMPORTANT ELEMENTS OF A DATA NETWORK

Our previous discussion brings us neatly to summarise the four main considerations affecting the design and operation of a data network (Figure 4.1):

(1) The computer code adopted to represent numbers, characters or diagrams in computer-readable format.

(2) The bitrate at which information is transferred.

(3) The transmission line or medium itself and the hardware interface (ie cables, sockets, voltages, etc) for connecting the end of the line (in the jargon, the *data circuit terminating equipment* or DCE) to the computer (the so-called data terminal equipment or DTE).

(4) The protocol.

It is these four factors which have the main influence on the correct performance of the data network.

If any of the factors (1)–(4) are not properly designed into the network initially, then the network may either not work, run erratically or run more slowly than required. Should these problems arise, the network needs re-design. During operation, on the other hand, problems generally only arise with the transmission line (3) or the protocol (4). Line breaks (eg cut cables) are the most common cause of failure nowadays, although electronic components (ie 'hardware') also fail from time to time. When protocol problems arise these are usually the hardest to diagnose and fix, since they are nearly always caused by incorrectly written software. It may be difficult to determine whether the software to blame is in one of the end computers

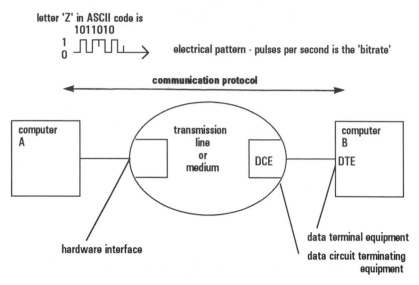

Figure 4.1   The basic elements of a data network

or in the network itself. Specialist equipment is needed for the diagnosis (data traces and protocol analysers).

We need to consider different types of transmission lines and protocols a little more closely, since these are the hardest part of network design. The next two sections cover these topics.

DATA TRANSMISSION LINES AND NETWORKS—TYPES AND CONNECTION INTERFACES

Short Distance Connections and Localised Networks

Over short distances, data transmission can be achieved between computers over direct cabling, provided some sort of appropriate software exists in both the computers to provide for the appropriate communications protocol. Thus for example, a computer terminal may be direct-wired to a host computer, and thus also, a Local Area Network or LAN of networked personal computers may be achieved.

In the case of a terminal directly connected to a 'host' computer (Figure 4.2), the protocol will most likely be one proprietary to the computer manufacturer (eg DEC, UNISYS, IBM, Siemens, etc). The cabling type required will also be specified by the manufacturer, as will be the maximum workable length of the cable (typically 10–100 m).

Local Area Networks (LANs)

In a LAN, the protocol is achieved by special 'network software' loaded onto each of the PCs. (Well known types of LAN network software are made by specialist software companies. Perhaps the best-selling softwares are IBM Token Ring (Netbios), Novell Netware and Banyan Vines).

If you read a technical book about LANs, you will learn about their evolution, and about the different basic topologies which are available (particularly *Ethernet* (IEEE 802.3) and *Token Ring* (IEEE 802.5)—see Figure 4.3). Experts tend to be emotive about which of the basic types— Ethernet or Token Ring—is technically superior, but for most users there is little to choose between them. Token Rings perform better than Ethernets at near full capacity or during overload but can be more costly to install— especially when only a small number of users are involved. The choice usually comes down to the recommendation of a user's computer supplier, since hardware and software of a particular computer or application type may have been developed with one or other type of LAN in mind.

Token Ring was developed by IBM, and is common amongst IBM personal computer users and as the network used in conjunction with the IBM AS/400 mid-range computers. Initially developed in a 4 Mbit/s form (which had the benefit of being able to be run on either coaxial cable or 'simple twisted pair' telephone cable) it has subsequently also become available in a 16 Mbit/s version. The 16 Mbit/s version is more expensive and demands the IBM cabling system, but may be crucial for ensuring acceptable network transmission speed for some of the graphical and image processing or large data-file transfer software which is beginning to appear.

In practice, the setting up of a LAN is easily planned—following the

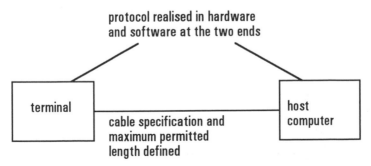

Figure 4.2   Direct cabling data connection

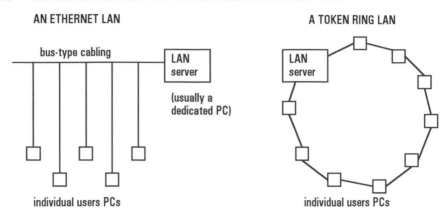

Figure 4.3   Main types of local area networks

software manufacturer's recommendations about suitable cabling and network layout (including number and location of the user stations on the LAN, and their maximum cabling distance separation). Typical maximum cabling length of the 'bus' or 'ring' is 200 m. Once the cabling is installed, including appropriate *hubs*, socket points and baluns (electronics in the socket which give the circuit the correct electrical characteristics), the software installation is usually straightforward, following a menu-driven process. However, the process can be laborious and may be best left to the supplier to perform.

Campus and Metropolitan Area Networks (MANs)

LANs are already the standard means of communication between desktop PCs and *server* workstations—so much so that the interconnection of multiple LAN networks is often necessary in large buildings and on campus sites. The problem of large sites is that the physical cable length constraints of LANs and the maximum number of users may be exceeded. Two basic methods are available for interconnecting LANs. They are bridging (performed by bridges) and routing (performed by routers). There are also bridge-routers or brouters which are a combination of the two.

Bridges are relatively simple and cheap devices which are suitable for connecting a relatively small number of networks together on the same site. In complex networks, and particularly in networks spanning a large number of geographically dispersed sites, they are unsuitable. They can cause network congestion of the individual LANs (slowing them up unacceptably), since they in effect broadcast all information to everyone. They can be very difficult to manage.

Routers are much more expensive than bridges but are the only manageable solution for complex networks spanning multiple feeder LANs. The router has more intelligence than a bridge and decides to which other destination router any particular information file should be sent. In this way, the load on the long distance network is kept to a minimum. Routers have evolved with the idea that they can be connected together in very large numbers—automatically conversing with one another to determine how individual messages can be delivered across a widespread network of multiple LANs. However, even routers have their limits, and, though the idea is that a limitless worldwide network of LANs connected via routers could be built, in practice the technology is not up to it yet. Extensive router networks need to be carefully designed taking due account of traffic patterns and of circular route possibilities—particulary at times when the network is heavily loaded.

Within the campus or metropolitan area a number of very high speed network technologies are available for interconnection of the bridges or routers. These are:

- FDDI (Fibre Distributed Data Interface) or FDDI-2
- MAN (either SMDS or DQDB).

FDDI is a 100 Mbit/s backbone technology for interconnecting local area networks using campus fibre. The second generation, FDDI-2, will include a capability for handling live voice conversation and video telephony over the same network.

MAN refers to a new extended breed of local area network technology, but one in which high speed data is not confined to a single office or building but may extend across whole cities or metropolitan areas. At the 140 Mbit/s speed of some early versions, a very large computer file (of say 1 Mbyte) could be retrieved from a distant source in well under a second. The most important technique in this bracket is a system called DQDB (Dual Queue Dual Bus). This is also marketed as SMDS (Switched Multimegabit Digital Service).

Wide Area Networks (WANs)

For long distance interconnection of computers or LANs a further category of technologies is important—the so-called WAN or *wide area network* technologies. There are multiple WAN technologies available. These range from techniques which simply provide point-to-point 'transparent' capacity to more sophisticated techniques which aim to optimise the use of the costly long transmission lines. Fortunately the techniques used to interface these

NOTES: DTE = Data Terminal Equipment (the user's computer)
                DCE = Data Circuit Terminating Equipment (a box of electronics
                      at the end of the main part of the circuit, usually located
                      in the user's premises, near his computer)

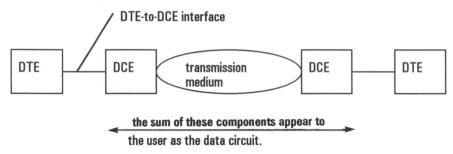

Figure 4.4   The structure of a long distance data circuit

technologies to the computer itself are standardised, so the choice between techniques can be weighted more towards economics and be less concerned with technical constraint. We discuss here the characteristics of the main network technology types, and the interfacing arrangements for connecting the computer. All long distance data transmission lines conform to the model illustrated in Figure 4.4.

We will review here the purpose and nature of the DCE, and the different types that are available to allow different types of transmission media to be used. We then go on to describe the DTE-to-DCE interface, and again the different types that are available and why.

The DCE is the equipment that is said to 'terminate' a data circuit. The DCE performs three main functions:

(1) It provides for a standard socket interface (the DTE/DCE interface) allowing easy connection of the user's computer equipment (the DTE) to the line. The interface specification includes the physical size and shape of the connecting plug and socket, the definition of the purpose of each electrical lead and, of course, electrical voltage and current levels. It will also usually state the maximum cabling distance between DTE and DCE.

(2) It converts the signal received from the DTE into an electrical format suitable for reliable accurate transmission across the transmission medium, and the distant DCE converts this signal back.

(3) It provides for clocking—ie it sets the rate at which bits will be conveyed

over the DTE/DCE connection—so that the two devices work in synchronism.

Where the circuit is a long distance one provided by the telephone company, the DCE is often provided by the telephone company at the same time. But in cases where the telephone company provides 'wires-only' circuits (eg the USA), the user will have to provide his own DCE.

Standard Computer-to-Dataline (DTE-to-DCE) Interfaces

There are a number of defined standard interfaces for connecting computer equipment (DTE) to datalines via DCE, as compared below. Generally it is not particularly important which interface is used, and indeed it is possible for the two DTEs (one at each end of the line) to use different interfaces.

The choice is usually determined by which interface is supported by both DCE and DTE. There tend to be national standards. Thus in the United States V35 is common, while in Germany, Deutsche Telekom only offers V24 and X21 for digital leaselines.

The most common DTE-to-DCE interfaces are compared in the following table:

Data interface	Physical connection	Bitrates typically supported	Maximum distance distance from DTE to DCE	Remarks
V24 (RS232C)	25-pin D-type connector	1200 bit/s, 2400 bit/s, 4800 bit/s, 9.6 kbit/s, 14.4 kbit/s, 19.2 kbit/s.	About 20 m	Electrical interface defined in CCITT V28. In consequence some people refer to V24/V28. V24 is equivalent to the American RS232C.
V35 or V36	—	1.2 kbit/s up to 512 kbit/s, but more common at rates above 64 kbit/s.	10–20 m	Commonly used in the USA on digital leaseline DCE.

| X21 | 15-pin D-type connector | Most commonly used at 48 kbit/s and 64 kbit/s, but theroretically possible at any rate. | 100 m | Commonly available on digital leasline DCE. The electrical part of the specification is also known as V11 or X27. Sometimes people refer to X21 but mean only the V11 part. |
| G703 | – | 64 kbit/s and multiples thereof, up to 2048 kbit/s (2Mbit/s). | About 10 m | Used when a high capacity line (eg 2Mbit/s) is broken down into individual 64kbit/s 'tributaries'. |

The connection of a DTE to a DCE can be relatively straightforward, but let me warn you of the following potential frustrations:

(1) It is sometimes possible in the communications software of a computer to define it as a DCE (rather than as a DTE). When configured as a DCE, there is some extra flexibility in the ability to connect to other types of device, but, configured as a DCE, it most likely will not work properly with the DCE at the end of the transmission line.

(2) Just because the cable fits in the sockets provided in the DTE and the DCE does not mean it is the right cable. (Sometimes, some of the wires in cables are cross-wired to different pins in the plug for good reason). To avoid potential problems arising from cabling, use the supplied cable. Check with your supplier if you think you have a problem.

Main Types of DCE and Dataline Available

Different types of DCE allow different types of transmission media to be used for the main part of a long distance dataline. The table below compares some of the commonly available types:

DCE Type	Transmission medium type	Remarks
Modem	Analogue (2-wire or 4-wire) transmission line. (The line may be a dedicated point-to-point leaseline. Alternatively, modems may also be used to create temporary datapaths across the ordinary telephone network.)	Most commonly available as units for 'desktop' use, maybe 25 cm × 20 cm × 5 cm in size. It is nowadays also possible to purchase modems already built-in to laptop computers. This facilitates, for example, remote access to the office email system.
Digital line DCEs	Digital (4-wire or optical fibre) transmission line.	DCEs may be known under various names: DSU (data service unit), CSU (channel service unit), NTU (network terminating unit), LTU (line terminating unit). Usually a device similar in size to a modem.
Multiplexer	A high bandwidth analogue circuit or high speed digital circuit.	A multiplexer allows for a number of devices at both ends of a dataline to share the same line, by dividing up a high capacity line.
Statistical multiplexer	Medium capacity anaolgue or digital line.	A statistical multiplexer makes higher use of the main part of the transmission line, by allocating capacity to numerous DTEs on an as-needed basis. So, for example, when one DTE is 'quiet' for just an instant, another DTE may be using the line. The key reason for the use of statistical multiplexers is as dataline concentrators—reducing the number of leaselines needed.
Radio modem	Radio transmission line.	May be an analog or digital device, allowing data to be transmitted across a radio medium.
Satellite modem	Satellite transmission medium.	A satellite modem may be either an analogue or a digital device. Users should beware of the longer time required for transmission via satellite—this can cause problems in both voice and data transmission.

PROTOCOLS FOR WIDE AREA DATA NETWORK COMMUNICATION

Protocols provide the etiquette of computer conversation—allowing computer machines to understand one another and communicate reliably. Perhaps the most important role of the protocol is to set up and clear communications paths and communications sessions, and ensure that throughout the session, the 'listening' computer is ready before the 'talking' computer speaks. Where, for example, the sending computer is able to shower the receiving computer with more work to process than it could handle, information could be lost if it were not for the protocol performing *flow control*.

Other functions performed by sophisticated protocols are the error checking (and sometimes correction) of received information, the interpretation of warning or problem messages, and the redirection of messages to avoid failures within the network.

Perhaps the simplest example of a protocol is one which works between a so-called dumb computer terminal and the computer's central processing unit or CPU. Computer users will be familiar with how they must press the ⟨return⟩ key before the message is sent to the CPU. Fairly quickly thereafter, the CPU responds as appropriate. In this case, the ⟨return⟩ key is being used as a *forwarding character*, signalling that the message is ready for transmission to the CPU. While the terminal buffer is waiting for the forwarding character (ie while the command is still incomplete), the characters typed so far are merely stored.

Another common feature of a dumb terminal protocol is the use of *echo*. The use of *echo* helps to ensure that there is not a cabling problem. When *echo* is in use, characters are not transmitted from the keyboard immediately to the user's terminal screen. Instead they are sent to the terminal buffer, from where an *echo* of the typed character is immediately returned to the screen. In this way, during periods of heavy communication between terminals and the computer, or during periods of line failure, the text appearing on the terminal screen may lag behind that actually already typed. This has two beneficial effects: one, it lets the user know there is a problem; two, it slows up his continued typing, so easing the problem.

Protocols and the networks that they are used with differ greatly—giving a wide scope of choice to meet a given data communications need. Some protocols are designed for use on point-to-point circuits (essentially directly connecting two computers) while others are especially designed to interact with switched data networks, giving the flexibility to connect one computer to many other different ones for maybe many simultaneous communication sessions. Thus, for example, a single connection from a host computer to a network might simultaneously serve the communications needs of a number of remote terminals all separately connected to the network.

The following table summarises the protocols most commonly used in wide area data networks and explains their use:

Protocol	Common usage
Asynchronous protocol	Most commonly used in connecting outlying computer terminals to their mainframe computers. Most asynchronous protocols are 'proprietary' to a particular computer manufacturer, but all are relatively simple and share similar principles. Asynchronous protocols usually need to be used on a point-to-point basis, although some networks allow such connections to be established on a switched basis — only for the duration needed. The Triple-X (X3/X28/X29) service associated with X25 packet switched networks provides for switched asynchronous connections. When using the Triple-X service, the PAD (packet assembler/disassembler) must be correctly configured. (The correct forwarding character must be set, echo must be on or off, and so on.)
X25	X25 is the most widely used protocol on packet switched networks. It allows for highly reliable and very flexible data communications. As an international standard, and one conforming to the *Open Systems Interconnection* (OSI) principles it is gaining popularity as the protocol of choice by equipment manufacturers and users alike.
Frame relay	Frame relay is a relatively new protocol — in effect developed from X25. The protocol is a 'stripped down' form of X25 — without some of the features of X25 which ensure reliable data delivery across relatively unreliable transmission media. The high reliability of today's digital networks make these features largely unnecessary. Frame relay places a higher responsibility on the communicating computers to ensure the integrity of data which they receive, but the benefit is a much higher transmission rate of data across the network than is possible with X25. This is important for bandwidth-hungry connections such as between LAN routers (discussed later in this chapter).
SNA (systems network architecture)	SNA was developed by the IBM company, and is used heavily by large industrial users of networked IBM mainframe computers. Some users find SNA between IBM computers to be preferable to the use of the X25 protocol, because they achieve better and faster overall network performance.

TCP/IP (transmission control protocol/internet protocol)	TCP/IP was developed by the American government as a protocol for the interconnection of large and widely-dispersed office networks and LANs. In particular, it allows for intercommunication of devices on different LANs (eg a PC on a LAN in a remote site could request information from a UNIX database held on a workstation in a central site). The protocol also supports widescale electronic mail operation between users on different networks.
Proprietary protocols	Many other proprietary protocols exist. Most of the main computer manufacturers have historically developed their own protocols. In addition, some small scale specialist communications manufacturers have designed equipment specific for given needs which also demand the development of specialised protocols.
Open Systems Interconnection (OSI) protocols	These are a new whole range of protocols being developed by the international standards bodies in response to the widespread demand amongst computer users and manufacturers for greater interconnectivity of different manufacturers' devices.

One thing is inevitable. The same protocol must always be adopted between the two end computers (unless some form of intermediate protocol conversion is employed).

Choosing which protocol to use, like other decisions, may be down to using the only one available. But should you have a choice I recommend that you use *open systems protocols* where possible (eg X25 or frame relay). You should, however, always consider that you will need to be able to call on specialist expertise should you need to resolve any protocol difficulties (though these are usually minimal once the network has been properly set up).

WHAT DOES ALL THIS MEAN IN PRACTICE?

In practice there are two main considerations in data networking:

(1) Designing and setting up the network.
(2) Operating it to a target quality standard thereafter.

Companies often have no choice but to develop a data network—it is forced upon them by their computer processing or company networking requirements.

The technical side of designing and setting up the network depends on four factors, as we have seen previously (using the right language code, determining an appropriate bitrate, choosing the network transmission medium and protocol). Often the code and the protocol are forced upon the data network designer by the prior choice of computer equipment. It then comes down to determining the transmission medium and bitrate. The questions to be considered in choosing an appropriate transmission medium are as follows:

(1) Which technologies are available?

(2) What specific communications needs do I have? (For example, switched network service or point-to-point ? Do I have a need for broadcasting?)

(3) What factors differentiate one network provider from another? (For example, cost, reliability, availability, promptness of attention to repairs, level of network congestion, supplier's prestige and credibility.)

The choice of bitrate may be limited by the end equipment to be connected, but sometimes the user may be able to choose between a number of pre-determined rates. So how should he choose? The simple rule is that the more information there is to be carried, the more users that will use the system and the faster the response time that they expect from it, so the higher the bitrate which is required. Working against the otherwise inevitable choice of the highest bitrate every time is the higher network cost associated with it.

The initial set-up of a data network can be quite prolonged and fraught with frustration. There are so many little things that must all be correctly set up before the network as a whole will work, and specialist help may be required to iron out, for example, initial protocol set-up errors.

Once data networks are working, they are generally reliable and easy to maintain. The most common cause of failure is the severing of the line — perhaps due to loss of power, or perhaps because a workman has accidentally upset the cable. Such faults can be easy and quick to troubleshoot.

A more serious problem, and one which is harder to resolve quickly is caused by the inevitable growth in demand for communication across a data network. As the demand reaches the capacity of the network, the whole thing slows down—like the traffic in a road traffic jam. The only solution is to provide more capacity—in short, to increase the available bitrate between congested endpoints. This may take a little time to organise (leased lines, for example, sometimes need at least a three month lead time). During the wait for more capacity, users are likely to become increasingly frustrated as the traffic grows further, and the situation deteriorates.

A still more difficult problem to resolve (but one which fortunately is rare) is an error in the protocol software which did not show up at network set-up. Such problems require specialists to resolve them.

WHAT YOU NEED TO DEFINE ABOUT YOUR DATA NETWORK NEEDS

The attributes we defined in Chapter 1 are helpful here, although as we illustrate below not all of them may be relevant:

Attribute	Context in data networking
Volume and frequency	1. How many, and which locations are to be connected to the network? 2. What is the type of each equipment to be connected, together with at least an estimate of the volume of data it needs to communicate and the time of day pattern of this communication? 3. How many simultaneous communications sessions must be supported by the network? 4. How and when are the sessions invoked by the computer applications? 5. Is connectivity required to 'external' organisations like customers and suppliers?
Image/quality level	1. Is a particular bitrate required by any of the applications used at a particular location? 2. What is the maximum allowable propagation time (ie delay) for data transmitted across the network during a session?
Information accuracy	Most data protocols nowadays see to it that data is not lost or corrupted by the network. However, if the line quality is poor, the error correction techniques used will considerably slow up the overall data rate. For this reason it is usual to quote a maximum allowable line Bit Error Rate (BER). This is a measure of line quality. A typical requirement is 10^{-9}.
Reliability/assurance of accuracy	1. How much redundancy (ie back-up connections and alternative routes) need to be designed into the network? (Some banks, for example, cannot afford to have their networks off the air, and so have a lot of duplicate capacity.) 2. In certain circumstances (generally only with very simple software and protocols) it may be possible that data can be lost if a receiving device is not switched on, is dormant, or is already busy on another task. A 'line splitter' used for broadcasting to multiple destinations, for example, is not reliable in delivering to all of them.
Confirmation of receipt	Most protocols expect many receipt acknowledgements during a single transmission—to check that previous data has been received before transmitting more. However,

these receipts are not visible to the end users.

Some software applications (for example, electronic mail) generate receipt confirmations for end users.

Answer/response needed

Some protocols automatically prompt for and initiate responses where needed. In addition, software applications may organise them (eg some electronic mail messages automatically require a reply).

Time available for delivery

1. Most data networks—for example, X25 packet networks and SNA networks—while technically being store and forward technologies, in practice usually deliver data within a second.

2. However, the network may not be (and may not need to be) connected all of the time. A dial-up network, used once per day overnight, might be a way of saving on call charges if only low volumes of data are involved and immediate delivery is not essential. (The use of dial-up networks from retail stores to headquarters how some retail chains report their daily sales results.)

3. On some types of Message Handling System (X400) data networks (including those used for EDI, Electronic Data Interchange) the delay between transmission and delivery may be several hours. You may need to check service delivery times with the supplier and monitor them from time to time, although the delay may be caused by the unreadiness of the recipient's equipment.

Broadcast

Broadcast can be achieved in a number of ways depending upon whether receipt assurance is needed:

1. The simplest and cheapest way is by using a line splitter. The same data is then fed simultaneously to a number of lines.

2. Satellite technology can also be used to a similar effect.

Both of these methods give simultaneous delivery to all destinations, but these may not all be ready to receive the information and confirmation of individual receipts may not be possible. If assurance of delivery is critical but same-time delivery is not then it may be better to use a specialised protocol—such as the X25 broadcast feature.

Collection of information

This might be needed, for example, for collecting sales reports from individual retail stores to the HQ computer centre.

Collecting data can be somewhat cumbersome and problematic if the number of outlying stations is high. Instead of allowing the outlying stations to call-in it may be better for the centre station to poll the remote stations. This avoids the possibility of all the outlying stations trying to call in at once.

Human contact

Videoconference applications, if they are to be run on a data network, place very stringent demands on the maximum allowable propogation delay and the fluctuation in this delay. This is likely to demand the use of one of the emerging high speed technologies—either frame relay or ATM (asynchronous transfer mode).

Record of communication

This will be up to the software application to organise.

Security of information

1. Is a physically separate network imperative to prevent unwanted intrusion on data privacy?
2. Is a radio or mobile network too insecure?
3. Will 'password' type software protections suffice?
4. Will a closed user group network suffice? (This is a network realised within the public network domain, in which only certain connections may call one another. It is realised in network software during set-up.)
5. Is data encryption necessary?

TYPICAL BUSINESS DATA NETWORKING NEEDS—AND SOME SOLUTIONS

It is not possible to describe and explain all the possible types of data networking needs which might exist within a business. Even less is it possible to cover comprehensively the multiplicity of solutions which are available in each case. Instead here we present a range of example networks and describe how the technology has been matched to the requirement, in order to give some ideas about how to satisfy your own data networking needs.

In the example of Figure 4.5, a direct leaseline connection is being used to connect a single terminal to a computer on a permanent basis. This solution is likely to be economic only where the leaseline length is very short (for example, across a site or to a building across the road) and where the usage of the terminal is particularly high.

In the example of Figure 4.6, the same basic arrangement of asynchronous terminal connected to a mainframe computer is shown, but this time the connection is made by means of a PAD (Packet Assembler/Disassembler—a device specially designed to convert certain types of data into a form suitable for carriage by an X25 packet switched network). This configuration is likely

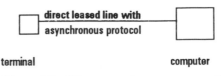

terminal   computer

Figure 4.5   Direct asynchronous connection

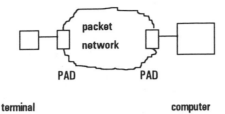

Figure 4.6   Asynchronous connection between the same end equipment as in Figure 4.5, but this time using X3/X28/X29 across a packet network

to be more economic when the terminal is relatively remote and/or little used. In particular, it may be a very economic way to run the terminal in a permanently 'live' state when actually little data is sent. (Perhaps the terminal is important as a monitoring device, and the keyboard is only used when something goes wrong, so that most of the time the terminal is dormant with no communication except occasional screen updates.) The PAD circuit is always ready to carry data, but the connection only needs to be active when something is typed on the keyboard. The rest of the time other users may be using the network capacity.

Figure 4.7 illustrates another variation on our first example. In this case there are a number of users in the same remote location, none of which have sufficient usage individually to justify a dedicated leaseline connection. However, by using a statistical multiplexer (stat-mux), a number of terminals may be connected back to the mainframe site using a shared leaseline.

Figure 4.8 shows a larger overall computer network—one in which a company has a central computer centre used by a number of outlying locations and offices. Since all the computer hardware and software was purchased from IBM, an SNA network is being used to connect remote cluster controllers back to the mainframe host computer via a Front End Processor (*FEP*).

In the example of Figure 4.9, the computing environment has grown still

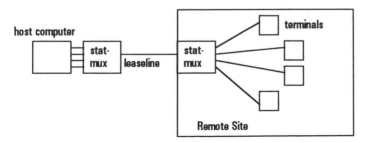

Figure 4.7   Remote terminals sharing a statistically multiplexed leaseline

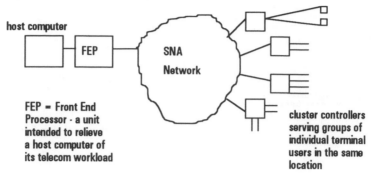

Figure 4.8 Major company with an IBM mainframe at a central site, and dispersed users

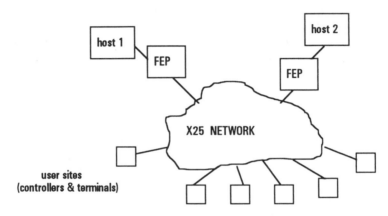

Figure 4.9 Even larger corporation than in Figure 4.8, with multiple computer centres—used for different applications

larger and more complicated. Now the company has a number of computer centres and a number of different mainframe computers which were supplied by different manufacturers. Individual computer users are served by different mainframe computers dependent upon their computing needs (for example, the ledger might be run on one mainframe computer, the sales and order processing on another, and the electronic mail on a third). In the configuration shown, an X25 network is being used because of its flexibility to switch data to different destinations, and also because of the interconnectivity requirement between computer systems of different types.

In the example of Figure 4.10, a dial-up network is being used as a cheap means of collecting sales information from individual outlets of a retail store chain. The sales made by each of the individual tills is fed to a personal computer in the store office, and, once a day, a datalink is set up across the telephone network for a few minutes in order to report the store results for the previous day. Profits and losses for the entire chain can thus be

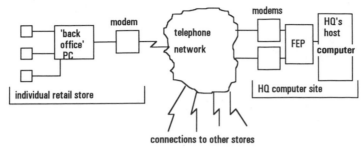

Figure 4.10 Reporting sales in a retail store chain

calculated next day, and depleted stocks in individual stores quickly replenished.

In the example of Figure 4.11, a satellite network is being exploited (because of its ability to broadcast the same information) to deliver live video reports and results from horse racing venues to thousands of betting shops. 'Satellite racing' is now common at betting shops in the United Kingdom.

Finally, Figure 4.12 illustrates the collection of financial information using a UNIX workstation and the subsequent distribution of this information to market traders in different locations using a TCP/IP network employing

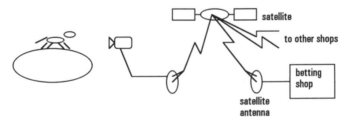

Figure 4.11 Broadcasting live horse racing to betting shops via satellite

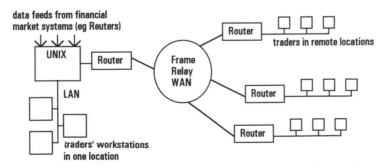

Figure 4.12 Distribution of latest financial market information to traders in a bank

routers and a wide area frame relay network. Compared with the Reuters terminal to which the trader may be accustomed, this LAN-based solution might give him more scope to process the data to his own needs (say, using spreadsheet software) rather than just view it on a screen.

TYPICAL COST STRUCTURES OF THE VARIOUS TYPES OF NETWORK

Most large scale private data networks have a cost which depends most critically upon telephone company charges for public data network or leased line services. In one of the international data networks I managed, the total costs were split in the ratio 5:4:1 respectively for the cost of leaselines, the cost of internal support personnel and the cost of equipment depreciation. The cost of the long distance lines was easily the greatest expense. The cost of the support personnel was also high because of the high staff numbers needed to maintain acceptable service to customers of our network.

Because the transmission costs of the network are so high, it is important to understand what the full cost of them will be. The table below therefore summarises the most usual tariffing practices for leaselines and public packet network service (X25). The table tries to alert you in particular to where you may find considerable hidden costs.

Data Network Service	Common charge structure	Possible hidden charges
National leaselines	Installation charge plus ongoing monthly or quarterly rental charge.	1. There may be a minimum contract period or a cancellation fee. 2. There may be reductions for longer contract periods. 3. In some countries there is a usage charge applied to leaselines—dependent upon how much data is transmitted.
International leaselines	Installation charge plus ongoing monthly or quarterly rental charge.	1. What is not always clear is that international leaselines are usually quoted as *half circuits*—so you have to add the costs quoted by the both of the national telephone companies involved (and probably will pay them in response to separate invoices).

| Telephone network usage for data- and circuit-switched data networks (including ISDN) | Installation charge plus rental charge for line connection and usage charge for periods of connection, dependent upon duration of calls and distance. | 1. Time-of-day dependent tariffs can vary greatly. Make sure the network is used at the cheaper rate periods whenever possible (eg overnight).

2. Watch out for data service surcharges on circuit switched networks like ISDN. |
| X25 packet network service | 1. Installation charge plus rental charge for line connection plus usage charges (which may be very complicated).
2. More recently, value added network suppliers and telephone companies have started merely to quote flat rate charges for entire networks, independent of the amount of usage. | 1. Where the tariff includes a usage element charge, be sure to understand all the elements and the likely effect on total cost. The Deutsche Telekom, in Germany, for example, charges on X25 network calls, three elements of usage charge:
* a charge dependent on the data volume;
* a call set-up charge;
* a call duration charge.
The cost of the call set-up and duration can amount to a large proportion of the bill, so remember them when comparing costs against a carrier who has only a volume dependent usage charge.
2. A flat rate charge has the benefit that you know your cost commitment, but in effect the carrier has charged you for the maximum usage of your network anyway. Compare this with what the cost would have been at his usage dependent tariff rates. |

THE PROBLEMS OF OPERATING DATA NETWORKS

Data networks require to be 'tuned' to their specific uses—adjusted to be an integral and well-matched part of a much larger computerised information processing facility. Problems in the computer programs can affect the correct operation of the data network and vice versa, so that problems which might at first appear to be network problems may not be so at all. A well-known frustration among the maintenance staff of public data networks is 'being blamed for a problem in the user's computer software'. The most frustrating part about this is that the customer may continue to blame the data network even after all checks on it have confirmed its health. But the customer may be right—the problem may be caused by the way the network interacts with the computer.

For example, the overall traffic on a packet network could be greater at some times of day than at others, thus causing slightly longer transmission times during these periods. This lengthened delay time may upset the computer program, which wittingly or unwittingly may have been written in such a way that the data transmission time is critical. Timing delays like this can upset many protocols and software—and of course neither the computer owner nor the network owner will accept that a problem exists in his equipment if it still appears to be working fine with everyone else.

There are no magic solutions, but the following advice may help a little:

(1)   Always be careful on long, international, and particularly satellite connections, as these are likely to suffer longer delay.

(2)   If possible, arrange the support personnel into a single troubleshooting team, so that there are not separate 'camps' for computer problems and data problems. If, however, you are only responsible for the network then try to ensure that your staff stick with the problem until fully resolved.

(3)   Be careful when adding new applications or traffic to the network, if this will add suddenly to the total traffic.

THE FUTURE OF COMPANY DATA NETWORKING

We finish with a short assessment of how data networking will evolve over the coming years—what new services and technologies seem likely to become available and what opportunities these might offer.

The main underlying trends which will affect the future of data networking are:

• the uses of and demands on data networking;
• the scale and scope of data networks;
• the services made available by specialist suppliers;
• the improving technology and consequently the costs.

The computer industry is rapidly moving towards windows-based software (eg OS/2, Unix and Windows), towards more personal computing and towards applications which use more graphical and image-based material. This will reduce greatly the relative importance of centralised computer processing bureaux and, instead, more widespread networks of individual computer workstations will appear. The nature of the image-based applications and the ever increasing demand for high performance and quick response will necessitate high capacity data networks with very high interconnectivity.

Business will continue to increase its dependence upon computing technology, and connectivity will grow to include not only internal business computer users, but, more importantly, customers and suppliers will also be much more strongly bonded by wide-reaching networks.

It will become increasingly less economic for private internal data network solutions to meet all the needs of the companies that they serve, and more Value Added Network Service (VANS) suppliers will appear to meet whole industrial market needs (eg the finance/broking industry, the retail market and its associated distribution channels, the motor industry, the travel industry and so on). This will continue the trend towards outsourced networks which commenced in the late 1980s. Increasing deregulation around the world will assist this trend.

In consequence of all these factors, technology will evolve towards meeting ever higher capacity needs, and ever wider geographical scale. Technologies which have already started to appear and which look of importance are as follows:

ISDN (Integrated Services Digital Network). ISDN is an evolution of the telephone nework—making it digital and capable of supporting integrated voice and data services. For carriage of data it will initially be important for applications needing short duration periods of 64 kbit/s connection (perhaps simple image applications or personal computer file transfers). In addition, ISDN may be important as a means of making networks more reliable—allowing 64 kbit/s leased lines to be backed-up on switched connections. We discuss ISDN more in Chapter 6.

Frame relay. This is a high speed packet data network technology that will carry data at very high transmission speeds. The high speeds enable large files to be transferred very quickly and network response times, even to short messages, to be much improved. Frame relay is a possible technology contender for the next generation public data networks—when X25 packet networks become obsolete.

ATM (asynchronous transfer mode) and cell relay. These are very high bandwidth technologies which have been designed especially to cope efficiently with so-called multimedia—networks combining simultaneously voice, data and video.

Mobile data networks. Public mobile data networks have already appeared in some countries. They are very efficient in their use of radio bandwidth, and bring a whole gamut of new business possibilities. For freight transport companies they open a new era. Not only can the goods manifests be checked at loading and delivery, their location during transit may also be monitored. And furthermore, new traffic report services available to motorists will help advise of the current road traffic situation—and of possible hold-ups to be avoided. Such a service is already available to motorists in the UK—advising motorists of hold-ups on the motorway system immediately around the London area.

Some express mail courier companies have recently impressed their customers with their ability to report the in-transit-status or delivery confirmation of parcels or consignments. Some even offer their customers the facility of a special data terminal to track their packages directly for themselves. This is certainly an attractive approach for providing extra customer service while simultaneously reducing query handling costs.

5

Message Networking Services

When people talk of the future of business in terms of 'networked companies' and 'networked offices', just what do they mean? Yards and yards of cabling around the building and always a technician taking up the floor or removing the ceiling tiles to install a bit more? Maybe. To me, however, it conjures visions of a more dynamic type of business organisation—one much more closely networked to its customers and to its suppliers, much more flexible in its working style, and more adaptive in reorganising its human resources to meet short term demands. This chapter discusses in particular the opportunities and technologies for improved inter-personal communication

MANAGEMENT STYLE AND THE 'CULTURE' OF PERSONAL COMMUNICATION

To meet the challenge of achieving the 'networked company' I see a need for revolution in the style of management. Much flatter management hierarchies and much less rigid reporting lines need to emerge, as does a higher level of trust in employees. Employees need to be expected to use greater initiative, enabling companies to be more adaptive to individualised needs. Short lifetime working groups may need to be pulled together quickly to see the task through. Staff will need to be more mobile—more able to go where the job dictates. There must be far fewer people locked-up in the corporate office and many more focusing on individual customers and new business needs.

Facilitating this revolution will be a new era of networking—one geared to direct communications between individuals rather than between the fixed offices and desks that they used to occupy. We have spoken already in Chapter 3 of the dawn of mobile telephony and of personal communications networks, but in addition to these, a plethora of new inter-personal messaging techniques are fast appearing.

Direct conversation and videotelephony will remain very important for dealing with sensitive customer service matters, for human team building and for negotiation. But at a lower level of human interaction some new specialist messaging technologies will take over general enquiries, information sharing and task coordination. We consider them technology by technology, explaining their strengths and weaknesses and how best to exploit them:

- Telex
- Facsimile (fax or telefax)
- Videotext
- Electronic mail
- Voicemail
- EDI (electronic data interchange)
- Radiopaging.

TELEX

Overall Assessment: A good method for conveying short text messages to international destinations which are difficult to reach by phone. The legal nature of telexes may also be important. Overall, however, telex is on its way out—supplanted by newer technologies.

Telex used to be the prime means for transmitting text. The answerback code generated at the receiving end is an invaluable means of confirming receipt at the correct destination. The legally recognised nature of this receipt confirmation has enabled telex for many years to be used to communicate legally binding documents—such as orders and contracts.

However, since the mid 1980s telex has been falling rapidly from fashion—pushed out by the growth in use of facsimile. Not only is facsimile easier to use but for messages of anything longer than just a few words, it is usually much cheaper.

Telex remains an important means of reliable communication to remote offices in countries where the telephone system is poor—so making it hard to get through on the fax. Despite this, its long term demise seems certain.

Figure 5.1   Telex—the answerback

FACSIMILE

Overall Assessment: A popular and convenient method for conveying text and diagrams. However, it is more expensive than many people realise, and its unreliability is often overlooked. Fine for ad hoc use, but should be supplanted with more robust messaging networks (such as electronic mail) between groups of very active users.

Basic Facsimile Service

Facsimile was the communications success story of the 1980s. The technology has proved very easy for users to understand and employ—and has found a thriving market amongst businessmen keen to share their documents and memos. The price of fax machines dropped quickly, so that it is now not only affordable, but common for companies not to have only one central fax machine, but many—scattered around individual employees' desks.

The technology has also adapted well to the personal computer environment. It is now possible to purchase fax cards for direct installation in PCs which do away with the need for a normal fax machine. Such cards can be added to office LANs, so making for easy sending of faxes by all the office employees. However, a normal fax machine may still be needed for receiving incoming messages.

While fax may be easier and cheaper to use than telex, it is nonetheless quite expensive—a fact which escapes many people. The high costs accrue from the approximately 45 seconds per page of telephone time (for a message sent long distance in the UK during the business day this equates to about 10p (20cent) per page—for an interational delivery to the United States this cost is about 50p ($1) per page). There is also a high cost associated with the special thermal paper (typically 5p (10cent) per sheet) needed by most receiving machines.

Thus a 20 page message sent across the Atlantic might cost around $22. This might still be cheaper and quicker than an overnight express courier, but a much longer message may not be.

As frequent users will know, long messages can be very tiresome in delivery—sometimes requiring several attempts. The first few attempts may be just in vain—finding the receiving fax machine to be busy. But even when it becomes free, it may take several more repeats to ensure that all the pages are delivered in a readable state. In the interim, phonecalls between secretaries may be needed to ascertain exactly how many pages have been received.

Broadcasting messages by fax is a particularly boring occupation, since the original often has to be fed through the sending machine many times—once per recipient. Some, more expensive, machines have appeared which claim to overcome this problem, but they need to be carefully attended to ensure that all the programmed fax numbers have been reached.

The trouble is that fax machines cannot be left to operate reliably on their own. Quite apart from the problems of them not getting through, and the possibility that several pages of the message may be drawn through the scanner simultaneously, there are also the annoying instances when another employee comes along and removes your fax (before its transmission) and replaces it with his own, which of course he believes to be more important.

Even once received, messages often end up scattered all over the floor, and some get lost in the internal mail before they reach their intended recipient. The print copy can also be poor, though newer equipment, and especially new generation *group 4* machines will in future give a much higher transmission and print quality.

The overall cost for fax transmission, the time for correct transmission and the full frustration I have never measured.

Value Added Facsimile Network Services

Against the background of difficulty in sending facsimile messages, a number of Value Added Network suppliers have seen the opportunity to launch services specifically aimed at addressing some of the problems. Let us also review the potential benefits and continuing risks of these services.

There have been two main influences which have driven the development of facsimile Value Added Network Services (VANS). The first was the concern of telephone companies for the degrading effect upon telephone networks that heavy facsimile traffic can have. (Some of the cost reductions possible on telephone networks rely on technologies which are able to compress voice signals. They do not cope very well with fax traffic and indeed may cause unacceptable degradation.) The second influence has been the recognition of the customer difficulties with the basic fax service (as we have previously discussed). The telephone companies have seen both these factors as an opportunity to introduce premium-priced services. Meanwhile, the equipment manufacturers have been developing specialised fax forwarding equipment for installation on company premises which is also designed to overcome some of these problems.

There are specialised equipments and fax VANS which cater extremely well with the needs of customers who wish to broadcast faxes to many destinations, or who wish to make cost savings on heavy international fax traffic or traffic to specific destinations. Overall, however, today's offerings still fall short of solving some of the main frustrations.

The fundamental difficulty with fax is its reliance on the telephone network and its technology—including a significant dose of human involvement required to ensure the successful delivery of each message. The very human-ness of the service is simultaneously its main strength and weakness!

Guaranteed delivery and confirmation of delivery are extremely difficult to achieve within any fixed timescale. But meanwhile the user perception is that delivery is immediate. Before actually attempting transmission, you tend to overlook the possibility of an endlessly busy receiving machine, and also those occasions when the recipient's paper runs out, or that you might have the telephone number to hand but not the fax number. Of course, if you stay with the job until you can confirm by telephone that the message is safely received, then you can achieve peace of mind—and this is probably how many users and secretaries operate today. However, could a machine be developed to achieve all these tasks with the same level of tenacity? Unfortunately the answer so far has been no.

Fundamental Difficulties with Automated Fax Services

There are four fundamental difficulties yet to be overcome by automated fax services:

(1) What to do when the telephone number given is wrong.
(2) Inability to guarantee delivery within a fixed timescale.
(3) Inability to confirm correct receipt at the intended destination.
(4) The resulting user concern arising from the lack of certainty.

While a machine might be capable of determining that the number dialled is wrong—that it is not for a fax machine but instead for a normal telephone—it has no recourse to alternative action without further human input (ie another telephone number).

If the dialled number is always busy, a fax VAN could keep calling* until finally it finds the recipient free, but there is no guarantee when this might be. So forget your peace of mind when relying on a fax VAN which proudly proclaims 'guaranteed delivery in 30 minutes'. The result still risks being overtime, being advised with too little time left (say, at 25 minutes that 'delivery has not been achieved yet'), or not being delivered at all (though you may be advised of this). This seems about the best that today's technology can achieve!

While it is possible to program facsimile machines to respond with their telephone number when receiving messages, this cannot be considered absolute proof of receipt at the correct destination. New telephone networks allow users to divert their calls to other numbers, at which a fax machine could be returning a false identification. So don't commit your most sensitive documents!

Positioning Fax as a Business Communication Tool

There is no doubt about the popularity of fax amongst ordinary business employees, and it would be difficult to eliminate. Nonetheless, as a company telecommunications manager I would personally prefer to put long term investment into alternative messaging networks, particularly electronic mail, rather than spend too much time in the short term developing a specialist facsimile network and trying to pretend to users that we have overcome the fundamental problems.

Short term, try to use a VANS supplier in order to save costs and avoid capital investment. The supplier may also prove to be a valuable assistant in assessing the full expense of your company's facsimile usage—and identifying the main users—by collating all the costs on one invoice. This in turn may help you to direct your initial electronic mail investment.

VIDEOTEXT

Overall Assessment: A must for travel companies in the UK, and otherwise interesting for information service companies in France. Beyond that, prob-

---

*Actually international standards now outlaw continuous reattempts. The problem is that continuous repeat attempts to the *wrong* telephone number can cause extreme annoyance to its owner if it keeps ringing in error. There is now therefore a maximum number of allowed reattempts.

ably already surpassed by information services specifically aimed at the personal computer user (eg public electronic mail services).

Videotext was originally developed as a low cost means of enabling its users to access information held in a public database. It never quite achieved the success it might have deserved and has now been overtaken by subsequent technology.

The idea of videotext is that, using a low cost terminal in the form of a small television, a customer could make a phonecall to a public central database, where he could access all sorts of pages of information which he could then have displayed on his screen. Thus, for example, he might access tomorrow's weather forecast, current flight arrival information, information about financial markets, about holiday offers, about dating services or whatever.

Videotext has found favour in only two markets that I am aware of. The United Kingdom travel industry uses it widely as a means for holiday companies to advise travel agents of available itineraries—and to book tickets on behalf of clients. And it has become the standard method for telephone directory enquiries in France, where the Minitel terminals were given away free by France Telecom. They justified doing this initially by the cost savings brought about by fewer telephone enquiries to human operators, but nowadays they also make money from the telephone calls which customers are making to the other information services which have subsequently sprung up. One such service, a dating service, is world renowned.

Apart from the above uses, videotext (also known variously as Prestel, Minitel and Bildschirmtext, BTX or Datex-J) is relatively little known. The sudden boom in the number of personal computers seems likely to scotch the chance of success for any possible revival attempts.

ELECTRONIC MAIL (EMAIL)

Overall Assessment: An exceptionally powerful and reliable means of business message communication, not only between internal employees but also offering the prospect of much closer contact with customers and better coordination of suppliers. It will form the basis of Electronic Data Interchange (EDI) between companies and so bring about a new era of more dynamic business.

What It Is and How It Works

Electronic mail is a reliable means of message communication between human users equipped with computer terminals or personal computers.

Large tracts of text, and diagrams too, can be quickly delivered across great geographical distances and, if necessary, printed to a high quality paper format using local computer printing resources.

To send a message, a user simply calls up the electronic mail software on his terminal. This will prompt him for the name of the user he wishes to send his message to, and for the names of any individuals to whom the message is to be copied. Having filled in this information, it will prompt him to type the main text of the message. He may be asked if he wishes to add (electronically) any other attachments (say a document previously created using his word processor).

Once the message is ready for sending, the system will prompt the user to confirm his readiness, and may ask him to set a priority rating for the message, to decide whether confirmation of receipt is required and to set time and date for delivery (if not to be immediate). Once these questions are answered the message (referred to in the vernacular as an electronic mail or an email) is timestamped and delivered electronically and more-or-less instantly into the mailboxes of all the intended recipients.

Should any of the recipients be at their computer terminals at the time of the message receipt in their mailbox, then they will be advised of its arrival (typically by a beeping noise and a short message displayed at the bottom of their terminal screen). They have the option to read the message immediately or later, depending on how they regard its priority.

Those recipients who are not concurrently using their terminals will be advised of the message next time they log on.

As messages are read, confirmation is returned to the sender (if required). Each recipient has the choice to reply to the message, forward it, file it, print it out, amend it or delete it.

The Advantages and Disadvantages

Messages are delivered extremely quickly with no possibility of getting lost or only part delivered. There is reliable confirmation not only of receipt but of users having read their messages. Broadcasting of messages is very quickly and easily achieved. Editing of text and returning or forwarding the amended version can be achieved with minimal re-typing. Messages can be filed and quickly retrieved later.

Messages can be posted for exactly timed delivery, and can be prioritised according to the urgency with which they need to be dealt with.

Furthermore, in the best designed and networked systems, users are able to check their electronic mailboxes even when they are away from their normal offices—either by using somebody else's terminal or perhaps by dialling into the system using a portable laptop computer from a hotel room.

The full power of electronic mail, however, will not be achieved until the whole staff of a company and all its customers and suppliers are connected to the network. Electronic mail should be the technology of choice for receiving customer orders, for laying-off orders on suppliers and for general communication, because it is potentially the most reliable, most accurate and quickest means of communication, overcoming all the barriers of geography and time zones.

Companies that have successfully introduced electronic mail have observed a beneficial change in the whole culture of how they do business. Questions and responses have been much quicker and more direct, messages have been much shorter and less formal—they have been typed by the managers rather than by their secretaries. Workgroups composed of members in widespread locations have evolved, and it is possible to draw together new teams for previously impossible tasks.

But there are also drawbacks.

From a user perspective, while delivery of the message to the recipient's mailbox is carried out within an instant, there is no guarantee that the recipient himself will read the message and so receive the information within any given timescale. This problem arises especially at the start—before the company as a whole adopts the whole culture of email into the way in which it works. A user who does not switch his terminal on (perhaps because he is apprehensive of its use) can devalue the system for all other users—who then learn not to trust it.

The high capital cost of installing an electronic mail system into an entire company can be a problem. But a slower rate of investment is not the solution, since that makes it much harder to achieve the full benefits. With only a few personnel connected to the system, electronic mail only allows them to communicate with some (rather than all) of their colleagues. In consequence, they have to continue using existing methods for contacting other staff. This negates some of the copying benefits of the system and indeed may even increase their workload.

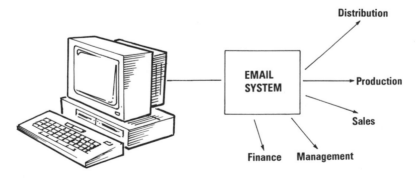

Figure 5.2 Quick and simultaneous broadcast by email

Introducing Email to your Company

Ideally *everyone* in a company should be put on an electronic mail system at the same time, for this generates for each of the new users unlimited bounds of communication using the new technique.

However, recognising that the costs of equipping the entire company may be prohibitive, and accepting the likely desire of management to see results first from a smaller scale field trial, what advice can be given about who to add first?

The simple advice is to ensure that users are added in groups—sets of people who have existing frequent contact in their day-to-day work as they carry out some common business function.

It may be appropriate to try email in a given department (for example, personnel or marketing), but more success may be achieved by adding all the individuals involved in a specific main business task (for example, customer order processing). In this case, we might take some individuals in sales and sales support, some in finance, some in customer service and some in operations—perhaps the team which together serve a particular geographic area.

Seed the trial by considering in particular which current communications within the workgroup could be transferred immediately to email, and include some ideas on how to do it (perhaps even some standard message layouts or 'electronic forms') at a training session to launch the trial.

Stress the need for individuals to be vigilant about checking their mailbox, even though for the first few days or weeks they may not often find messages for them. Encourage use of the system not only for formal business processes (such as customer orders) but also for quick notes and questions between personnel—even if they are only social ones.

Be prepared to add further individuals to the trial at short notice—perhaps in response to a request from the trial participants to add further colleagues. This helps to reinforce the value of the system to users who have already discovered individualised uses, and gains sponsorship for new users.

Make sure at least one senior manager is involved in the trial—and choose one if possible who has experienced email before and is already a strong advocate. His or her enthusiasm will help to ensure the trial gets a fair chance.

One potential work group for the trial is, of course, the group of the most senior managers. Many companies have introduced email this way. But while these individuals may form a valuable or even inevitable trial group, they carry with them a number of very high risks:

(1) Senior managers can be tempted to leave the job of the email to their secretaries—this largely defeats the object—removing the spontaneity of quick response and informality.

(2) By choosing the most senior group, you ensure that probably only the chairman has his communications needs met anywhere like in full. He will be the only one able to communicate with all his first line reports and regular company contacts. All the other senior managers will suffer the frustration of having to convert to a different communications medium when forwarding information to their own staff reports.

(3) Unless the group as a whole is committed, or the chairman adamant for success, the trial may not be given the fair chance it is due, and patience to continue it may soon be lost.

(4) The failure of a senior management trial kills any possibility of adjustment and re-launch.

Email as a Message Toolkit for the Mobile Employee

Given not only the explosion in numbers of personal computers used by companies, but also the wide availability of portable laptop computers, there is an opportunity for extending the use of the electronic mail 'network' for keeping in contact with managers even when they are working out of the office.

Little more is needed than a modem built into each laptop PC, a special version of the electronic mail software for remote PC use, and a telephone number to call at the computer centre where messages are stored.

The special PC software (available from most of the main electronic mail software suppliers) is designed to prevent longer telephone connection to the central computer than absolutely necessary. It makes a call to the central system to deliver and receive messages, and then clears the line until another message is ready to send. Messages can meanwhile be prepared 'offline' on the PC.

The use of a freephone telephone number might reduce any possible inconvenience that users might encounter in calling the centre, and reduce the cost of telephone mark-ups on calls made from hotel bedrooms.

Another possible part of an international electronic mail toolkit would be a telephone plug adapter device—enabling the traveller to plug his PC into a hotel telephone socket in any country in the world.

EDI—Communicating via Email to your Customers and Suppliers

Electronic communication between companies is already being practised, and will become an increasingly important way of doing business. Many supermarket chains demand that their suppliers accept orders for goods electronically. This allows the supermarkets to place orders daily, and so maintain their shelf stocks (and value of stocks) at a minimum level—just enough to meet one day's sales. It helps the supplier too—for he can ensure that products on the shelves with his name on are fresher and closer to their best. Perhaps even more important for the supplier, it allows him to bill for the goods more quickly.

Electronic communication is critical to speeding up the cycle of business—making suppliers along the full length of the 'supply chain' far more reactive to the demands of the market.

Making electronic communication possible between companies in this way is a new range of technical standards for Electronic Data Interchange (EDI). The most important of these new standards is the United Nations' EDIFACT (Electronic Data Interchange for Administration, Commerce and Transport), but there are a number of regional, national and industry-specific standards which may be of importance to your company.

Most of the EDI standards are available in ready-packaged software solutions for integration into a company's existing computer and electronic mail set-up. Value Added Network Services have appeared in several countries which specialise in encouraging the connection of all companies within a specific industry segment. In this way the VAN creates a community of interest for communication with external customers and suppliers (for example, within the retailing industry or the car manufacturing industry).

Setting up the technology for EDI is actually quite straightforward. But

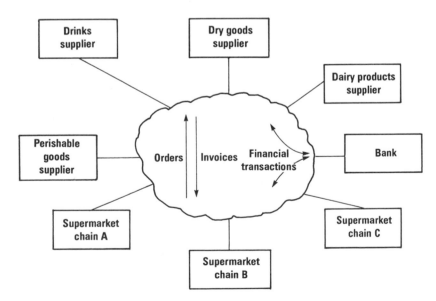

Figure 5.3   An EDI 'community of interest'

what is much harder—and what the company therefore needs to be prepared for—is the adjustment of its culture and practices to an entirely new way of business requiring much quicker reaction to customer demands. It's a risk, but one worth taking, for it offers the prospect of getting one step ahead of the competition. In fact, if your competition is successful with EDI, you will not survive without it.

Choosing the Right Email System

The costs of running an electronic mail system include:

- The cost of a personal computer or computer terminal for every user;

- The cost of equipping these with relevant software;

- The cost of a computer storage point of some kind for acting as the central post office for messages in transit through the network; and

- The cost of the telecom network linking all the parts together.

There are a number of slightly different technological and service approaches possible for electronic mail. We shall consider three of these:

(1) Private mainframe service—serving only company employees from a large central computer.

(2) Private LAN-based service—serving only company employees but using a network of LAN mail servers rather than a central mainframe.

(3) Public electronic mail service—renting mailboxes from a specialist VANS supplier, who ideally is already providing mailboxes to customers and suppliers within your community of interest.

Where a company already has an existing mainframe computer which is used by a large proportion of employees it may be possible to instigate a private electronic mail service at marginal additional cost—requiring only the purchase of system software and disk memory capacity for message storage. It should be recognised, however, that electronic mail can be extremely heavy in its use of both CPU time and storage space, particularly if the electronic mail software includes a word processor, and generally mainframes are not a cheap solution. For these reasons it may be worth either installing a dedicated minicomputer for the use of the electronic mail (so that other existing computer programs are not unduly upset), or instead finding a package which relies more heavily on the processing power of users' individual PCs for preparing messages, and if necessary uses the mainframe only as the central post office.

In choosing the system for your company, you should consider beforehand the range of facilities that you will need (not all electronic mail is the same). I suggest you buy a system with at least the following minimum set of features:

Important electronic mail 'feature'	Minimum requirement that should be sought
Directory of addresses	This ensures correct spelling of the recognised addressee name.
TO and COPY at letterhead	Systems which have only TO lists can be confusing, since they do not identify the main recipient.
Message prioritisation	Possibility for sender to allocate some sort of priority, so that receiver may deal with his incoming messages in priority order.
Sender's copy	Not all systems retain a copy for the sender. This can be annoying.
Incoming message alert	Immediate alert to recipient if he is currently using his computer.
List of incoming messages	Some systems merely present all the messages one after another in the order of receipt, which may mean a receiver has to read many messages before he sees the one of relevance.
Receipt confirmation	Ideally it should be possible to receive a receipt confirmation if requested at time of sending. Receipts for critical messages may be important, but can be a nuisance if received for all messages.
Message status enquiry	It is useful for senders to be able to query the system about a particular message to determine which recipients have read a particular message, and when they did so.
Message reply facility	It is helpful to have a reply facility which automatically inserts the addressee list—either as 'to sender only' or to 'same copy list'.
Message forwarding facility	Enables easy onward copying of messages to further recipients.
Electronic filing	To enable easy later retrieval and/or editing.
PC software compatibility	It is helpful for users to be able to transfer their word processor documents or spreadsheets directly into the mail system. Windows software compatibility may be important.
X400 and X500 compatibility	(Explained later in this chapter.) This makes for easier interconnection with electronic mail systems of other companies.

In running the system, you need to find a way to encourage users not to store too many messages for long periods of time, since this can quickly generate a huge cost due to the need for extensive computer disk storage. It may be worth setting a maximum time limit before automatic deletion. Before this period expires, users who wish to retain messages need to find some means of archive (paper print out, or downloading to their personal computer).

The alternative to a mainframe-based electronic mail system is a LAN-based one. There are many to choose from, and the same selection criteria as above should be used. A LAN-based system will be more suitable in companies who are already heavily LAN-oriented or who seem likely to evolve that way. As we noted previously, the inherent advantage of a LAN-based system is that each user employs his own personal computer for preparing messages, and so does not disturb other users by making heavy demands on the central computer.

Where a LAN-based system is used, individual LANs in different offices or buildings need to use the same software, and be linked together. This should be done using the standard equipments for LAN interconnection, ie Bridges and routers (see Figure 5.4). Bridges are straightforward and cheap where only a small number of LANs are to be interconnected, but generally I would recommend the use of a router for more complicated networks. Overgrown LAN interconnect networks without routers can get very complicated to manage and can become overloaded and consequently very slow in operation. Be very careful in designing the routes and storage points for electronic mail messages across complex networks—it is very easy unwittingly to establish circular routing paths which may cause both considerable delay to message delivery and substantial network congestion in the meanwhile.

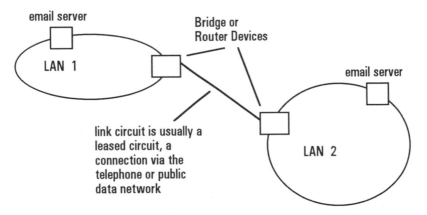

Figure 5.4   Standard methods of LAN interconnection

Finally, let us consider the use of public email services provided by specialist VANS suppliers. The principle benefits of these services are:

- No capital start-up cost.

- Immediate communications community of interest with other existing customers (particulary useful if the VANS supplier specialises in your industry segment and so already has your main suppliers and customers connected).

- May be better suited for field staff or managers who travel frequently, since public email services often offer localised access points in many (even international) locations. Thus the manager visiting the United Kingdom from his home office in the United States may only need to call a London telephone number to check his mailbox, rather than make a call back to the corporate HQ.

- System upgraded with new features more frequently, at someone else's capital cost.

The main disadvantage of public email services is that over a longer course of time they are inevitably more expensive than the private alternative.

X400 and X500 Capabilities

X400 (or more correctly CCITT Recommendation X400) is the name given to a series of technical specifications called the Message Handling System (MHS). In the context of electronic mail, these specifications provide an important means for connecting electronic mail systems of different type (ie from different manufacturers). It is therefore important to ensure that your chosen software has this capability. The connection to other companies' systems may either be made directly using the X400 protocols or it may be more effective instead to install an X400 connection to a public email service and use this as the transit point for reaching other networks (see Figure 5.5). As with LAN-based email solutions, be very careful in the design of an X400 email network—preferably making it very hierarchical in order to avoid the message delays and network congestion which can result from unplanned circular routing patterns.

But while X400 provides for the technical interconnection, it will be impossible for your users to send messages to users of other systems unless the correct electronic mail name and address is known. These may be known by word of mouth for a small number of users on other systems and so

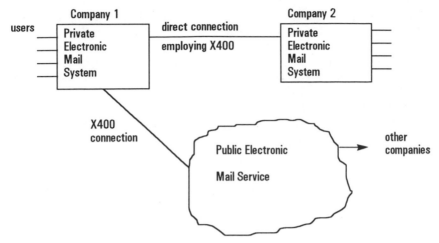

Figure 5.5   Alternative means for connecting to other companies' electronic mail users

intercommunication can commence. For wider range of contacts, some sort of equivalent of the telephone directory service is needed. This is provided by means of the X500 Directory Service. The principle of X500 is that anyone in the world connected to an electronic mail system can be addressed by any other electronic mail user using a simple 'normal' naming convention. Thus my address might be:

Country:	Germany
ADMD (administrative domain):	DBPT (Deutsche Bundespost Telekom)
PRMD (private management domain):	Deutsche Bank
Department:	Deutsche Gesellschaft für Netzwerkdienste
Location:	Eschborn
Name:	Clark, Martin

The ADMDs are distributed by CCITT on an international basis. ADMD owners (generally the national telephone companies and other VANS providers) then distribute PRMDs within their ADMD. Individual PRMD owners (typically companies, like the Deutsche Bank) then are free to design addresses within the remaining address space.

If you get faced with designing your own, try to make it as simple, logical and explicit as possible. If I were the second Martin Clark to join the Bank in Eschborn I would not like to have the address Clark, Martin P or Clark, Martin Philip while the first user retained the address Clark, Martin. Why not? Having already suffered this one, I can tell you that message senders, on finding that Clark, Martin is an accepted name, assume that this is the correct addressee name without checking the subsequent initial. The result is that the existing user gets all my mail. It's very annoying, I can assure you. How do you avoid it? You give every user at least two initials right from the start.

VOICEMAIL

Overall Assessment: An excellent messaging medium as the standard communications methodology amongst particular types of workgroups — especially field-based ones such as sales and field support.

We have previously covered (in Chapter 3) the subject of voicemail in some detail — explaining its strengths and weaknesses. It remains here only to make a direct comparison of voicemail and electronic mail, to help those readers who may feel confused about the relative strengths of the two technologies.

Voicemail is similar to electronic mail in its ability to serve the messaging requirements of work groups. And like electronic mail, it is generally only successful in cases where an employee can communicate with most of his usual contacts over the system. In this way, voicemail (or electronic mail) can become his standard method of communication. The particular voicemail system selected for a company, and the method by which it is introduced, should therefore follow similar criteria to those just discussed for electronic mail.

Some companies have introduced both voicemail and electronic mail, recognising their different strengths. For others, the annoyance of having to check two separate types of mailbox is enough to make them elect either for one system or the other.

The best voicemail systems signal to you when you have a message waiting by means of a lamp on the handset and/or a modified dial tone. Each time you pick up the phone you'll know if a message is waiting and there is no need to make a special enquiry call.

Voicemail tends to be most liked by field-based workgroups (eg service maintenance organisations, sales staffs, etc). For these individuals it offers the convenience of them being contactable in the field — using available and familiar telephone services — and avoiding the need to carry PCs. Electronic mail, on the other hand, is easier for office-based staff who may already be equipped with computing resources, who may wish to share computer file information and can be alerted immediately on screen when incoming messages arrive for them.

RADIOPAGING

Overall Assessment: The only reliable means of ensuring the immediate alerting and availability of critical staff. Cheap and effective.

Radiopagers are the 'bleeper' and message devices, usually about the size of a cigarette packet, which may be paged by callers—usually by a simple telephone call to a dedicated phone number. On dialling the paging number, there is usually a recorded message answer saying that 'Mr X is being paged, please hang up'. Alternatively, there may be an answer from a human operator at a central despatch bureau. The operator will take down a short text and then organise its delivery to the relevant recipient.

Wearers of radiopagers are advised of the incoming message usually by a bleeping sound, which is set off by radio signal to the pager—wherever it is. In the case of a message pager the bleep will be accompanied by a message of up to 80 characters. Having received the bleep, the wearer calls back to his base office from the nearest telephone to receive more information—perhaps even to speak with the person who has paged him.

Radiopaging is really the only reliable means of ensuring that key staff can be contacted in times of emergency. I would recommend the use of *Message pagers* where available and affordable, since they can be used to give the message recipient some indication of the nature of the emergency and its priority. This will help him determine the relative importance of the message relative to work he may already be involved in—with the minimum of disturbance to him.

6

The Company's Backbone Telecom Network

No matter how keen it may be to avoid investing in telecommunications assets, any company is bound to own some of what it uses, if only the cabling around the office. But how much and which technology should it be using? And how much should it privately own? These are the questions which this chapter attempts to answer.

HOW MUCH TELECOM EQUIPMENT SHOULD YOU OWN?

It is neither possible to answer in absolute numbers how much a company should spend on telecommunications nor to specify exactly which technologies should be in use. These are questions the company itself must answer, by review of its working practices—in particular by reviewing the communication channels between the different departments and the various different business functions as we have already discussed in Chapter 1.

THE COMPANY'S 'BACKBONE' NETWORK

Having helped the business determine which technologies are needed, the communications manager faces the job of deciding how best to implement the various networks that are needed to achieve maximum effectiveness and economy. The result of this process is what I refer to as the company's backbone network or as its communications infrastructure. Thus the infrastructure for a small company may comprise only a single telephone line rented from the local telephone company. Meanwhile, for a large multinational company (like an international bank) the backbone network may be a very extensive privately owned network of transmission links, switching equipment, dedicated buildings and operational staff the world over, interconnecting all the various branches of the bank on a permanent basis.

SMALL COMPANY NETWORKS

Given a small expenditure on telecommunications by smaller companies, there may be no other economic option than to rent nearly all their telecommunications services from the local telephone company. But while many telephone companies remain staid and inflexible, you can rely on them to come back—and to fix your problems eventually. In fact, the 'customer conscious' suppliers which are supposed to thrive in competitive markets have not brought improvement for all customers. Some customers have found that the local telephone company fitter, once a man to engage in a quick chat and one who would go out of his way to do the odd favour, has now been regimented to complete the job on his task card as fast as possible and get on to the next one.

You may consider the telephone company too expensive and you may be right, but at least you will not be committed to large capital investment and you probably will not need to recruit a specialist telecom manager to your own staff.

There is often no cancellation fee for terminating public network services that you no longer need. This may give you the scope to try technologies you might not otherwise have been able to afford.

For small companies, the quality and efficiency of the company's telecom infrastructure will rely entirely upon the development of a close working relationship with the management of the local or national telephone company. The customer will be entirely reliant on his supplier to recommend potential economies and to bring to notice any new services or technologies appropriate to their business. Such advice is obtainable even from the monopoly telephone companies, but it does take a little effort at the start to find the right commercial contact, and to develop the right human relationship.

BIG COMPANIES—THE PROS AND CONS OF PRIVATE NETWORKING

The greater expenditure on telecommunications by larger companies is usually accompanied by a great desire for minimised cost. This in turn is the single most important incentive for a company to consider installing its own private network.

The term private network is used to describe a network of privately owned telecommunications equipment installed on private company premises for private company use (typically for telephone calls between employees located in different premises, towns or even countries). However, the telecommunications cables and circuits which connect the different sites of a private network are, almost without exception, leased from the public telephone company. (See Figure 6.1—it is usually only the telephone companies who are licenced to dig up public roads and lay such cables.) There are cases where private companies are permitted to lay and own their own transmission links (cable or radio). These are usually on-site at large campuses (eg universities), or on private wayleave ground (as for example owned by the railways and electricity utility companies).

Private networks can be built to serve any type of network service (eg internal company telephone service, company voicemail service, computer data network, etc). The economy of private networks arises from achieving high usage of equipment or circuits which have been bought or leased for exclusive use.

Where cost savings in long distance telephone calls are the main objective of a private network, the cost comparison needs to be between the flat rate rental charge for a leased circuit and the volume dependent charge if the same calls were to be sent over the public telephone network. It is always the case that at a given threshold of usage between two specific points, a leased circuit will be cheaper. Typically the threshold duration at which the

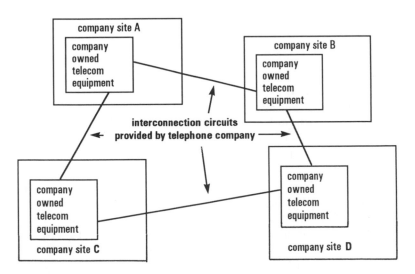

Figure 6.1 A typical private network

leased circuit becomes cheaper is around one to two hours usage per business day.

Private networks can be targetted specifically to handle company voice or data network needs, but generally most economy will be achieved by developing a transmission network that will carry both. How to do this is discussed in a later section in this chapter 'The case for and against combining voice and data on the same network'.

Apart from the potential cost savings, the other principal advantage of private networks is the ability it gives a company to control its own network—to determine its own service level (be it high or low) and to choose the most suitable technology. For some high-technology-minded companies, the adoption of new technologies by the large telephone companies is too slow. By installing a private network, they need not wait for all the technical difficulties of interconnecting the new technology with all the existing public networks to be solved. Thus the private network enables them to achieve the benefits of the latest technology (whatever they are) at the earliest opportunity.

Once a private network is installed, there is scope for all sorts of little enhancements and refinements—company-wide four-digit dialling scheme, 'ring back when free', centralised network control centre, and so on, to name but a few. But my advice is to avoid too much additional effort on 'frills'. What starts out as a 'nice to have feature' might become a major focus of operational effort, and start to obscure the real reason for having the network in the first place. After a few years of cost savings achieved by a private telephone network, the economics may suddenly change to favour using the public telephone network for all calls. Then you may be forced to decide how much premium you are prepared to pay for those frills—the unified company numbering scheme and the 'ring back' facility.

Private networks are common in countries where leased circuits are cheap relative to public network services. Historically they have been more common in certain European countries than in the United States. The United Kingdom in particular has had a high number of private networks. However, there is a growing trend around the world for much more competitively priced public services (ie lower public telephone call tariffs), while leased circuits are tending to get relatively more expensive. This has resulted from regulatory and market pressure on telephone companies to make their prices more cost-based.

My own expectation is for this to create and maintain a trend away from private networks. In their place I expect a wider range of more specialist Value Added Network Services (VANS) to appear—offered by operators who are bringing together communications 'communities of interest' (multiple companies in the same industrial sector). Because of their scale, VANS will yield cost economies that are possible only for the largest networks.

At the same time the ex-monopoly telephone companies will be introducing a wider range of new services, some specifically targetted to 're-capture' private network business. These new services are tailorable to individual company needs and much more competitively priced (ie telephone company profit margins will return to more reasonable levels). Examples which are already starting to appear are:

- Virtual Private Networks (VPNs). These aim to be a direct replacement for existing private telephone or data networks — enabling companies to give up their private network for the lower costs of the public services and yet not have to give up features like company numbering scheme, 'conference calling', 'ring back when free', and so on.

- Centrex. This public network service aims to remove the need for a company to have its own on-site telephone PBX.

Unless the ex-monopoly telephone companies allow their market shares or geographical coverages to disintegrate too far, their scale advantage over new competitors is simply too great for them to be seriously challenged on price. Only the most aggressive competitors with huge capital resources or niche-market VANS offerings will present real threat. For even at new customer sites in the most remote places, the old monopoly telephone companies inevitably already have a handful of existing customers, and a basis of existing equipment which can be updated at marginal cost.

WHAT ARE FACILITIES MANAGEMENT (FM) AND OUTSOURCING, AND SHOULD YOU USE THEM?

Facilities Management (FM) and Outsourcing suppliers provide:

- Specialist staff for planning, installation, maintenance and operation of the client's computer or telecom equipment (as needed).

- Equipment, floorspace and high grade computer operating environments (air-conditioning, power, uninterruptible power supply (UPS), etc) for installation of the client's own computer or telecom equipment.

In effect the client is able to run his own private computer system or private telecom network without the need for his own specialist techical planning and support personnel, computer rooms and operational staff. The prospect may be attractive for a company lacking sufficient accommodation or finance for a computer room or, alternatively, for a company wishing to avoid the costs and constraints of recruiting specialist staff outside of their core area of expertise. Even in cases where the company already has specialist sites and human resources in one country, it might lack the

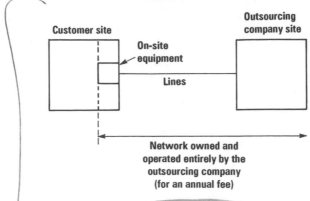

Figure 6.2   The principle of outsourcing

justification for these in another more remote area where there is a much lower level of business activity.

A number of cases can be cited of companies who have chosen to outsource their entire networks and computer operations (eg Kodak). By doing so these companies have sought to reduce their costs and to reduce their workforce to staff working on their core business.

Often the first step in large scale outsourcing has been to sell the existing network assets and operations personnel to a specialist networking company, which then sells back network services at a running cost ideally lower than the client had previously been paying. In cases where the client's cost can be reduced in this manner, it is because the network company has either been able to make economies by rationalisation, or because they have been able to make better use of the client's assets by selling further service to third parties.

Usually the client agrees to buy service from the outsourcing supplier for a given minimum period, after which he expects to renegotiate price and other terms at a competitive market rate—moving to another supplier if necessary. In so doing he aims to ensure ongoing minimum cost.

A spin-off benefit on the personnel side of outsourcing is that the specialist network staff now all work in a company whose core business is networking. This potentially leads to better management of the human resources, and better job prospects for the employees.

At the lower end of outsourcing, what is basically the same service is instead called facilities management. This is aimed at companies wishing to leave the operation of their remote sites to a network company, and at companies wanting an operations environment on a small, and otherwise uneconomic, scale.

A particular benefit which facilities management and outsourcing companies ought to be able to offer is low cost leased line capacity between their sites, and short lead-time establishment of new client circuits. This is

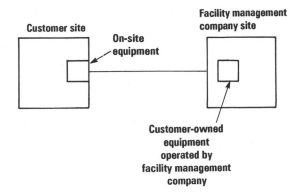

Figure 6.3 The principle of facilities management

achievable by the leasing of high capacity links between their own sites, and the use of these for multiple customers (in effect 'buying wholesale and selling retail').

My cautionary notes on outsourcing and facilities management are as follows. A big company selling its existing private network to an outsourcing supplier needs to consider how much control it will lose of its network (and/or computing resource). It may gain a better grip on the price, but will this be at the expense of reduced flexibility to meet its needs at short notice, and reduced responsiveness to resolving problems? Will it thereby lose any of its competitive edge? Perhaps a bank, for example, needs to consider communication as part of its core business.

For a company buying facilities management only on a small scale there is the additional risk of inadequate technical expertise on the part of the supplier. While many facilities management companies proclaim expertise as part of their product, this may not mean expertise in the particular equipment that a client wishes to have managed. They may themselves be reading the handbook for the first time. They obviously have personnel with general networking training and experience, and this may be all that is required in simple cases, but do not expect advice on, or operation of, the more sophisticated aspects of your equipment. Remember, your equipment may be one of a number that he has, and with which he may be only 75% familiar.

THE CASE FOR AND AGAINST COMBINING VOICE AND DATA ON THE SAME NETWORK

A question often asked is whether voice and data services (e.g. computer transmission, fax and any others) should share the same core network. The simple answer is that if services could share the same network in some way, then there will probably be some sort of cost saving. But the critical question is to what degree the services should be integrated, for it is possible for voice and data to share the same transmission without the services themselves being fully integrated. Let us consider the following analogy to explain the difference.

Two cooks share the same kitchen. One, Smith, is a pastry cook—he uses a lot of flour. The other, Jones, cures meat—he uses a lot of salt. Each cook *rents* three storage jars for their dry ingredient from a kitchen supplies company. (Why they rent them may puzzle us, but this is critical to our analogy.) They each have separate rental agreements. One day, Jones and Smith get together and strike on the brilliant idea of cancelling their separate agreements for rental of three jars each, and instead agree to make a joint agreement for rental of six jars—since the supplier gives a 40% discount for agreements of four jars or more.

Our example is exactly analogous to the shared use of high capacity telecom lines split up into individual segments dedicated to the use of either voice or data. Thus the six jars are analogous to a high capacity (lower unit cost) transmission line. This is then broken into two separate and independent segments—some is used just for voice (analogous to the flour) and some for the data (analogous to the salt). Thus by sharing the transmission medium (the high capacity line), cost savings accrue from bulk purchase.

One day, our cooks make another realisation. They realise that neither of them ever uses more than 2.5 jars—or five between them. So that if they mixed the flour and salt into the same jars they could save another jar and make a further 16% saving (from six to five jars). But to do this they would need very specialised equipment for sorting out the salt from the flour when one or other of them were needed for cooking.

This second example is analogous to a telecom network in which voice and data are *fully integrated* (as, for example, in the ISDN or integrated services digital network). The overall number of transmission links is reduced in integration by allowing voice calls to use free data circuits, and vice versa. The balance of voice and data usage of the transmission capacity will change minute by minute. The technique relies upon the fact that it is unlikely that all voice and all data circuits will be required at exactly the same time—but sophisticated technology is required to re-sort the calls back into voice and data types, just as the flour and salt needed to be separated.

Our analogy (though somewhat hypothetical) also illustrates the relative cost benefits:

• Sharing the underlying transmission network between voice and data (ie by adopting high capacity lines and using multiplexers to break them into segments for dedicated voice and data usage) is relatively easy and cheap to put into practice and achieves a relatively large cost saving—because for a given capacity, a few high capacity lines are cheaper than more lines of lower capacity (the bulk purchase effect).

- Integrating voice and data services on the other hand is technologically much more advanced and expensive to achieve, and may be without much further cost saving.

It is normal today for a company's telephone network capacity needs to far exceed those of its data network (a 64 kbit/s leased line would typically carry only one or two telephone calls, but just a few of these lines could keep an entire mainframe computer busy!). The expenditure ratio may be 10 times as much on voice as on data. Over the next few years this ratio will reduce as data grows more rapidly than voice, but for now the network needs to be optimised around voice needs. Meanwhile, it may be possible to get the data network as if for free. Take, for example, two large sites and the need for 25 private voice and three private data lines between them. Achieved on separate transmission networks this would require one 2 Mbit/s line (capacity 30 lines) for the voice circuits plus three extra data circuits. Sharing the 2 Mbit/s transmission line between voice and data (using a multiplexer) would obviate the need for the three extra data circuits, by using some of the spare 'voice' circuits (as in example 1 previously).

Alternatively, the entire network could have been turned over to an integrated services digital network (ISDN—as in example 2 previously), in which case the 2 Mbit/s line would still have been used, but now its entire capacity could dynamically have been reallocated between voice and data use (allowing 30 voice conversations at times of no data—or even 30 data circuits at times of no conversation). This might have some benefit from an operational viewpoint, but from a pure cost standpoint alone it may be that the ISDN network upgrade costs exceed the cost of the multiplexer we need in the case above.

Unless the PBXs are already ISDN compatible it is likely that the ISDN upgrade will be the more expensive. So just from a cost-saving point of view, ISDN may not be the best solution. However, the slight extra cost may be justified by some of the powerful new services made possible by ISDN. We discuss these a little later in this chapter.

SPECIFIC TECHNOLOGIES YOU SHOULD CONSIDER FOR THE BACKBONE

No matter whether a company is large or small, it needs to consider its transmission infrastructure needs. Below we consider a number of technologies and services which may be appropriate (depending upon overall network size).

Company Office Cabling

The right office cabling is perhaps the most important part of a company's telecom infrastructure. If you get it wrong you can spend a lot of time and money continuously changing it. Cabling is best installed in a structured manner, and then left alone. Continually taking up the office floor or removing the ceiling tiles not only causes annoyance to the office staff, it also disturbs existing cabling, and so creates difficult-to-find intermittent faults.

Many different types of structured cabling system are available nowadays. It is not important exactly which equipment is used, but you should try to ensure that:

- The type of cable installed can be used for as many different telecom uses as possible (ie the same type of cable for voice, data, video, etc), The cheapest cable—telephone twisted pair—is sufficient for telelphone, normal computer terminal lines, together with Ethernet and 4 Mbit/s Token Ring LANs. However, higher grade cable (eg IBM type 1) may be needed if you anticipate very high speed data networking (eg for 16 Mbit/s Token Ring or image applications on high speed computer workstations. Fibre optic cable should also be laid to the workplace if high speed data or video applications are foreseen during the lifetime of the cable system.

- The cable is laid in plentiful quantity to sockets nearby each potential desk location or other workplace (the amount should be higher than even your wildest estimate of foreseeable requirements, and at least four pairs per location—a digital telephone may require two pairs, a computer terminal at least one pair and a fax machine at least one).

- The cable should be physically laid in a star-formation back to a central patch panel, where different *logical* configurations (eg LAN ring or bus) can be created by making different cross connections.

- Standard, and robust, plug and socket technology is used—both in the individual work locations and at the patch panel. Poor connections are the hardest faults to find since they are often intermittent in nature. They are most likely to arise when cables are frequently re-patched.

In very large sites, it may be appropriate to use a separate structured cabling system for each building (or even each floor of a large building). Then the separate patch panels can be connected together using fibre cables and 'active' cabling components (ie equipment) which perform the multiplexing and signal conversion necessary between copper and fibre cable types.

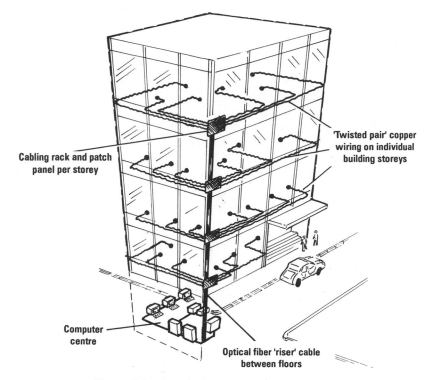

Figure 6.4   A typical structured wiring scheme

Leased Line Types—Analogue or Digital, Terrestrial or Satellite

Leased lines may enable costs to be reduced for transmission between locations which share particularly heavy traffic.

Leased lines will be either analogue or digital, and provided either by some sort of terrestrial means or by satellite. In general, analogue lines are disappearing in favour of digital ones. Where the choice does exist, the factors to be considered in selection are:

- Is your end equipment more suited to analogue or digital lines? (Data lines are more suited towards digital, while for an old company telephone system it may not matter.)

- What are the relative costs of analogue or digital lines?

- What is the relative *availability* of the two services (ie what percentage of the time can you expect them or need them to be working)?

For long distance and international lines you may have the choice as to whether the circuit is carried by terrestrial means (eg cable, optical fibre, radio) or by satellite. Either may be cheaper, dependent upon circumstances. Satellite services in general have a very high availability (they are less likely to be disturbed by shipping trawlers or road diggers). There are no hard and fast guidelines as to which will be better quality—generally this comes down to the quality of the supplier and its staff, not to the technology.

From a technological point of view, satellite can cause problems in voice and data networks. In a voice network it can add long pauses into conversation, and create an echo of the speaker's voice—returned 1 second later. With modern echo suppression or echo cancellation techniques, the echo can be eliminated, but in poorly designed or maintained networks this may not operate reliably. The 1 second pause after you finish speaking and before your correspondent appears to answer is scientifically unavoidable. This is offputting to telephone users. In a data network, the extra delay can cause insurmountable software problems for the end equipments.

Multiplexers

Multiplexers are the pieces of equipment which allow high capacity transmission lines to be broken down into smaller segments for use as individual voice or data channels. In general they are simple and relatively cheap devices of very reliable technology. Prices range typically from $1k to $40k dependent upon the level of sophistication required. Most units nowadays have the option of a network management workstation for remote reconfiguration of the multiplexer. This might allow individual circuits to be changed from voice to data use (say, overnight) without necessitating a visit to site. TDM or time division multiplexers are required for digital lines, while FDM or frequency division multiplexers are required for analogue lines.

Fast Packet and Frame Relay Switches

The fast packet and cell relay (ATM or asynchronous transfer mode) switching techniques are rapidly becoming the standard replacement technology for TDM or *circuit switched* multiplexers—allowing different types of services (eg voice, data and video) to share the same transmission media in a more economical and dynamically adjusted fashion than is possible with simple multiplexing. The techniques open the door for much faster data processing—larger file-handling capacity (as will be needed for new video and image applications) and much faster network-response times.

As a replacement for simple multiplexors these devices are of importance in private company networks. But the public network operators will also offer frame relay service as a foundation for high speed wide area data networking.

ISDN (Integrated Services Digital Network)

As we saw in the previous section of this chapter on 'The case for combining voice and data on the same network', you may not find a simple cost justification for integrating voice and data services—as might be achievable by an investment in a replacement network using ISDN technology. You may find the ISDN investment worthwhile nonetheless because of the need for one of the new services ISDN makes possible. We do not cover all of these new services in this chapter or even in this book, but for backbone infrastructural reasons the following services may be interesting:

- Dialled back-up (ie circuit restoration using a dialled connection) for 64 kbit/s digital leased lines—ensuring almost 100% availability of the data service carried by the circuit. (Alternatively this facility could be used to provide temporary 64kbit/s lines, if a leased circuit installation were delayed. Of course the ISDN line would have had to have been previously installed.)
- Low-cost, on-demand switched videotelephone service.

Hybrid Technologies

Hybrid technologies are capable of supporting simultaneously different types of telecommunication transmission (eg frame relay and TDM circuit switching). The *raison d'être* of hybrid technology is the economics to be gained from combining voice and data networks. But, rather than compromising either the voice or the data or both by attempting a poor full integration (as in ISDN where data is poorly catered for), hybrid systems instead allow for a coexistence of separately optimised voice and data channels.

The most sophisticated types of hybrid technology allow for dynamic sharing of bandwidth according to instantaneous demand. Thus, for example, a hybrid switch might allow voice and data users to share a 64 kbit/s long distance circuit. Up to three voice channels might use 16 kbit/s of bandwidth each, while whatever remaining bandwidth not currently in use

(16–48 kbit/s) could be used in parallel to provide data transmission using a frame relay protocol.

Hybrid technologies will be very important for private network operators in the run-up to full ATM availability. You particularly need to watch out for hybrid network technologies capable of very efficient private switched voice network solutions.

BUILDING THE NETWORK 'JUST IN TIME'

When a network is in a state of growth, considerable savings are possible by delaying expenditure on capital equipment items until the last moment possible. The manufacturing industry has recognised this for a long time, and named the principle *just-in-time* (JIT). In short, what we aim to do is provide no more circuits and switching equipment than are absolutely necessary to handle the very next day's traffic demand.

Telecom equipment nowadays lends itself very well to just-in-time provision. Most systems use only a small number of different electronic cards, and can be built up in a modular fashion—allowing operators to expand the size of their equipment almost at will—requiring perhaps only slide-in installation of an extra card and a simple software reconfiguration from the control console. Choosing the right modular equipment is not only a safeguard against unforeseen later growth needs, but also might delay some of the capital expenditure during a planned growth phase (because you don't have to buy so much at once).

However, while the equipment may lend itself to just-in-time provision, you may need to be a little more circumspect in relying upon the telephone companies to install new circuits or other network connections on time. The telephone companies have not proved to be the most punctual of suppliers! In my experience, the extra money spent on ordering the circuits a little early is handsomely rewarded by the avoidance of user complaint either when their service is delivered late or when the network as a whole becomes unbearably congested.

WHAT IS THE ROLE OF NETWORK MANAGEMENT?

Every network needs management. Problems need to be diagnosed and dealt with quickly, and in the longer term the trends in demand need to be monitored and new network capacity planned accordingly.

Network management staff should be judged on how well they manage the following network attributes:

- Network quality. Is the quality of transmission acceptable to users? Are the right services being offered?

- Network availability. For what the percentage of time during the month was the network available and in operational order? (You may only be interested in percentage of *working* hours available.)

- Network congestion. What is the percentage of calls not reaching ring tone or customer busy tone on a telephone network, or what is the transmission delay time of a data network?

- Network security. To what extent is there proper confidentiality of information passing across the network? What measures are available and when was the last audit? (There is more about this subject in a later chapter.)

- Faults. What is the total number of faults and what percentage are cleared within one working day? Also, how responsive do the end users think the operational network staff are?

- Invoice accuracy. Bills from telephone companies have been notoriously inaccurate and need occasional audit. Lines which have been cancelled may still be being charged.

- Network changes. How well are they planned to minimise outage time, how well are they carried out, and documented?

Most networking equipment today comes with its own sophisticated computer control console. Many of these have very clear graphical screens which are extremely effective in displaying the current status of the network (% operational, % congestion, and so on—perhaps displaying in red any failed components). They allow quick diagnosis of problems even by relatively untrained operators. Such equipment is critical to the basic management and operation of the network, and is often installed all together in a central control or network management centre—a grand term, but in some cases the network management centre need be no more than a single PC and a responsible operator (for a company PBX this one person might be the office receptionist and switchboard operator).

Over and above the separate control consoles for each individual type of equipment (eg modems, switches, etc), there is a current trend towards the development and installation of umbrella network management systems. These are consoles designed to control the separate equipment consoles. The idea is that a single operator will be able to understand and control all the separate network components—all from a single workstation.

Bringing control to a single workstation (as has already been largely achieved by systems like those widely using *simple network management protocol* or SNMP is only one step of an ongoing network management

Figure 6.5   Management of network congestion

revolution. The much more important step (and the one not yet fully achieved) is the development of *expert* network management systems capable of simplifying network control. Thus, for example, while it is already possible to load routing tables into each node of a complicated voice network from a single workstation, most of the hard work remains—working out the routing table which needs to be loaded at each switch and going through the manual action of loading it. The network management systems are not yet at the point of working out the routing tables from a simpler input by the operator. However, development continues, and this type of system is bound soon to be commonplace.

NETWORK COSTS AND HOW TO RECHARGE THEM

Networks rely upon the cooperation of individuals to share communications, means and costs. No-one would use the public telephone network if he had to pay for it all—it would be far too expensive. But because everyone else in the country subscribes to a line and pays for it, a customer has a wide choice of people he can call—for a moderate and affordable cost. Once a network has reached a certain critical size, and has become stable, its financing and its cost allocation becomes relatively straightforward. For smaller networks, particularly those which are either growing or shrinking, cost allocation is more problematic.

If all the costs incurred before the first customer were added were all charged to the first customer (or the first customer department using a new company telecom service), there would never be a first customer. Not only would he have no-one to call, it would cost him a fortune if he could do so. The problem is that just at the time when most money is needed, the network has least to offer. The usual ways around the problem are first to seek an initial critical mass of customers (or subscribing departments) so that, shared between them, both the service and the costs are acceptable. The second way is to accept the need for a fairly lengthy payback period on the project (ie by depreciating the capital cost over a longer period—say five years instead of three), thus lowering annual costs which must be recharged to customer departments. This second approach in effect makes the network department a risk venture business.

The risks are twofold: first that the equipment will not last the planned lifetime, second that customers find they no longer need the service and stop paying for it earlier than anticipated by the budget. In this case, there may be a temptation to recalculate costs—dividing by the number of remaining user departments, who in consequence all find their costs higher than expected. This can lead to a re-evaluation of the benefits in each individual department and, as more pull out, the problem gets worse.

The morals are simple. First recognise that any amount of private networking is in effect a business risk venture. Second, if possible try to avoid too-rigid direct division of telecom network costs between participating departments, as this may only distort an objective view of their real control of telecom usage. Instead think of the network as a business venture in its own right, that may make or lose money and affect the overall company results. If you don't like the risk, buy public telecom services instead.

If the company likes to recharge network costs to individual user departments, this should be done at a commercial price—one including a small overhead of profit. The profit is critical for supplying the investment for any new telecom services.

Consider, for example, a case where a network department installs a voicemail system. Let us imagine that the minimum size of equipment which may be installed would serve 100 users and that this can subsequently be outfitted at a marginal cost per user (typically a relatively cheap extra 'user port' card) for the needs of extra users up to a maximum of 400 users. To provide for the needs of more than 400 users a second system must be purchased. These 'step increases' in costs are common when purchasing or extending telecom equipment. The overall capital costs plotted against the number of users is then as shown in Figures 6.6 and 6.7.

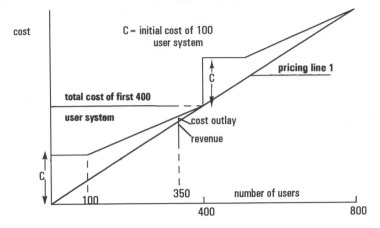

Figure 6.6   Pricing of voicemail system—option one

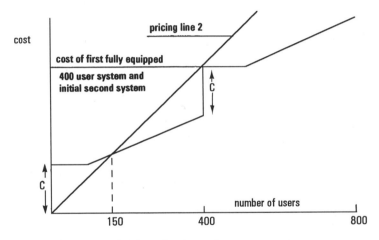

Figure 6.7   Pricing of voicemail system—option two

The question is what price to set for the initial users, assuming that the 400 limit of users will be relatively quickly surpassed? We discuss two possibilities, shown as 'pricing line 1' and 'pricing line 2' on the diagrams.

You may consider pricing line 1 (Figure 6.6) to be the fairest method of recharging users. This represents the exact division of the costs for a fully utilised (400 user) system between the 400 users. Using it, however, leads to almost certain problems. First, if the system is only ever subscribed by 350 users then there is a lot of money short (since the cost line on our diagram is above the price line (ie recouped costs) at this point. But worse still, if 401 users turn up then virtually the entire initial cost, C, of the second system will be in arrears.

A better pricing line to use is pricing line 2 (Figure 6.7). This divides the full cost of the first 400 user system and the costs of the initial part of the second system between the first 400 users. This has two benefits. First the venture is in profit after about 150 users are added to the first system. Second, there is enough money available to install the second system as soon as the need arises.

Pricing line 2 may not be justifiable for a very large number of users (say, more than 800) since it may generate unreasonable profits. Once the initial and second system purchases have been made and sufficient paying customers have been amassed, there will be scope for price discounts. Any new pricing formula, however, should have at least two things in common with pricing line 2:

(i)   the marginal revenues collected from any new customers should at least cover the next step increase in costs (when it becomes necessary);

(ii) individual users should be offered a competitive price, which by providing equipment and service of their own, they could not match.

THE PROSPECTS AND PITFALLS OF OFFERING SERVICE TO EXTERNAL CUSTOMERS

Some companies consider reselling their private network capacity to external companies in order to share some of the costs. This might be appropriate in trying to 'lock-in' a customer or get better access to a supplier. However, as a commercial prospect in its own right, such a service should not be entered into without full thought for the consequences.

As soon as external customers are added to a company's private telecom network, the need for running the network department as an independent business becomes practically unavoidable. Neither the existing customer (ie the company itself) nor any new customer will want to be treated as second priority. The new customer will expect confidential information passed to the network department not to be circulated into the company as a whole, and the old customer (the company who actually owns the network) may have to accept that they are no longer the priority that they were. They may not find this easy, or even acceptable.

It is easy to believe that network services could be resold to third parties from an extensive private network, but there are a lot of adjustments to make. First, the network does not actually connect to any other customer's sites, so there will be extra investment in infrastructure needed. Second, the organisation will need to be capable of providing adequate customer service and accurate invoicing. Third, the network company will have to commence independent marketing and sales activities if it is to secure its future.

Some companies' network departments have successfully made the transition: *Tymnet* (now British Telecom's Global Network Service) was originally the network of McDonnell Douglas and *AT&T Istel* was originally that of the British Leyland Motor Company. Both have become successful Value Added Network Service providers to other major companies.

TAKEOVERS AND JOINT VENTURES

For any number of reasons, network services companies and company telecom departments get faced with having to integrate networks of previously different history. This typically comes about because:

• The parent company makes a new business acquisition.

• The would-be VANS company needs to extend its geographical coverage by joint venture.

The most common problems in attempting to integrate networks are:

• Different cultures of the staff.

• Different basic technologies, leading to loss of some of the sophisticated features, also to duplication of expert staff for operations, maintenance and support.

• Difficulties, in particular, in integrating administrative functions. Combining the billing systems, for example, can be a complicated and expensive task.

Before taking on the task of integration you need to be clear about what needs to be achieved. Possible objectives are:

• Increased network reach and/or market share of both networks in the short term.

• Network infrastructure and support costs savings in the medium term, or improved service for the same cost.

• The acquisition of larger customers.

• The alignment of technology to achieve the economies of scale necessary for more sophisticated services.

Having thought about these, you need a comprehensive plan and enough time to achieve it. The process is rarely simple.

HAVING THE GUTS TO KILL YOUR 'PRIZED POSSESSION' WHEN THE TIME HAS COME

All business products have a life cycle, and there comes a time when the product has little more useful value. Once it reaches this stage it is time to replace it with something else, killing the old product as quickly as possible to avoid the likely mounting costs of supporting it. Telecom networks are no different from other products in this regard.

How recently did you question whether the company's private network was still economic in its current form?

7

International Networking

International telecom networks present a particular challenge because of the marked differences in the national telecom markets. Companies developing or using international networks face additional problems over and above those met within national networks.

INTERNATIONAL DIFFERENCES

The problems of international networking arise from:

- Differences in regulation — what is legally permitted.
- Lack of true international network suppliers.
- Differences in culture and organisation of the national telephone suppliers and of the users.
- Difference in the range of public network services which are available.
- Differences in technology and technical standards.
- Differences in prices and pricing structures.

The development of an international network demands understanding of the different national situations and flexibility in approach. What may be the best technological solution in one country may not be in another.

In this chapter we review the main differences, and offer some ideas on how to manage in each of the countries, and how to overcome the main problems.

WHY THINGS WILL CHANGE

The root cause of most of the problems is the rigid monopoly structure which existed in most countries until at least the mid-1980s. The state-owned national telephone companies could afford to operate in a relatively inefficent and isolationist manner, offering only the services to their customers that they chose to. Many of them, in nationalistic style, developed their own technical standards, and in some cases even the equipment itself. What has resulted is a wide variety of technologies, some good, some bad, but without the international market necessary to further the good ones and weed out the bad. Even the basic telephone socket has a different variant in each country—but for whose benefit?

Fortunately the situation is bound to change. The worldwide deregulation of telecommunications, particularly the introduction of competition, will see to that. New competitors compete on price and service to attract customers from the established networks. They do this by bringing new ideas and keeping costs low. The big telecom equipment manufacturers are keen to support the newcomers, particularly if they bring with them a new market opportunity as well. By growing, the large manufacturers are able to reap greater economies of scale in manufacture, and an expanded base of capital for supporting their research and development. They become more efficient and their products become better value—not only for the national telephone companies but also for individual customers. The purely national manufacturers gradually will be unable to compete except in niche or protected markets.

The network operators too are growing and becoming more international. By extending their geographical network coverage outside their traditional home countries, network operators become more able to protect their existing base of customers, particularly large multinational conglomerates with extensive international needs. The carriers become more attractive to their customers against other international competition and simultaneously generate new revenue from established clients. For the multinational customer, the more homogeneous service and the greater control of his supplier brought about by his greater purchasing leverage can be a significant advantage. So the market is evolving for international network services—a trend towards greater customer orientation and technology standardization.

As with the manufacturers, it is the largest international network operators who will be able to reap the greatest economies of scale. These are the operators who should be able to develop their market share most quickly in a new era of end-to-end international network services. Already the process of rationalisation of the big national telephone companies has started and some big international players are beginning to appear. I do not expect them all to survive.

DEREGULATION

Deregulation is the misleading term which in the telecommunications world has come to be associated with the re-structuring of the national laws to introduce competition in telecommunications networks and services. National governments have differed greatly in the speed with which they have embraced deregulation—differing on their views of its benefits and disadvantages.

Some governments have regarded deregulation as critical to stimulating investment in the telecommunications infrastructure of their countries, and are urged on by economists who seem in no doubt that good telecommunications infrastructure is a critical precursor to national economic growth. These are the governments who have been quick to privatise their state-owned telecom monopoly and to encourage widescale competition. Other governments have been much more cautious in deregulation and have jealously guarded the monopoly telephone company as a national asset.

According to how far they have deregulated their telecommunications market, countries may be split into five categories:

(1) Those which maintain a monopoly market.

(2) Those which maintain a monopoly for network services but which have freed the market for purchase by users of their own telephones and other *customer premises equipment* (CPE).

(3) Those which allow competing Value Added Network Services (VANS).

(4) Those which allow competition in voice services.

(5) Those with an open telecommunications market.

These categories are the visible signs of deregulation, and many countries on the face of it have deregulated to stage 3 or stage 4. Customers may not have noticed so much progress, however, for undoubtedly the most critical factor determining the ease of doing business in any particular national telecom market is the will of the government to support change through competition.

No matter what the law may say, what is possible in practice is determined by government will, and some have introduced telecom competition only as the result of external international pressure. In these countries theory often differs markedly from practice. For example, it may in theory be legally possible for new operators to compete against the monopoly carrier, but the administrative process for registering as a competitor may not exist, may be incredibly slow, or may expose commercially sensitive business plans of new operators to their main competitor—the monopoly carrier. Other barriers can be technical ones such as failure to release technical specifications which allow interconnection with the monopoly carrier, or—much more simply—failure to release telephone numbers which would allow the new carrier to offer services to customers. Barriers such as these have existed in several countries which have claimed (for political reasons) to have deregulated.

In those countries where competition has been introduced there has been a marked reduction in telecom prices, an increase in the range of available products and an improvement in customer service. In Japan, three years of competition brought the long distance call tariffs down by more than 50%. In the United States, the initial years of competition had similar effects, and today the largest corporations pay less than 10 cents per minute for long distance calls which may link East and West coasts. This is a far cry from the $1 per minute rates (or higher) charged for some long distance calls in Western Europe.

Which is the easiest telecom market to do business in as a customer? Undoubtedly the United States. Competition there is very hot, service very good and profit margins low—you can get a lot of value for money. Suppliers there recognise that the customer's satisfaction comes first—and much more flexibility is demonstrated in their eagerness to win business. They realise that their future rests in their own performance, and they are ever keen to improve it.

Many of the suppliers in Western Europe and on the Pacific Rim are becoming much easier to deal with, but some of the old lethargy is still about. Prices may be keener and the marketing brochures glossier, but is the leased line installed more quickly than it used to be? Unfortunately at some of the root levels the stirring effects of competition have yet to be felt!

THE EMERGING INTERNATIONAL VALUE ADDED NETWORK SUPPLIERS

Capitalising upon the wave of deregulation which is sweeping the world, a new breed of network supplier is emerging—one which provides its customers with end-to-end networking solutions.

In the main these new operators are able to offer only Value Added Network Services (VANS), typically offering managed packet data networking service. Examples of emerging global networking companies are:

- British Telecom's *Syncordia* and *Global Network Service*;

- *Sprint International's* data network service;

- *AT&T's Istel's* international data network service;

- *Infonet's* international data network service.

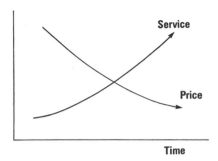

Figure 7.1   The effect of competition

What the new operators are offering are more global networks, designed and built to cover huge regions with seamless services, technology and operating practices. They will begin to displace the old international networks made up of a patchwork of independent national sub-networks.

Not only because of their scale, but also because of their independence from other companies' operational practices and pricing policies, these new international VANS should be more flexible in meeting business customer needs, and more competitive on price, but you need to recognise their limits.

In most countries where these new network suppliers operate, they do so as VANS providers. Under such provisions they are usually permitted to build large scale networks using their own telecom switching equipment installed in their own premises, and to connect direct customers to these networks. But in doing so they must use lines provided by the monopoly carrier or an equivalent licensed transmission network provider.

The obligation to use lines provided by the monopoly telephone company is probably the biggest constraint faced by VANS operators. The cost of leasing the lines can be a huge proportion of the VANS operator's costs (typically around 50%). Fluctuations in price are uncontrollable, and worse still, in many cases are unpredictable as well. This hinders forward network planning, and can make the VANS price volatile as well. In addition, the unpredictability of service can be a problem.

If you order a connection to a VANS, the supplier is bound to lay-off an order to the national telephone company for a leased circuit from your premises to his nearest VANS exchange. He may not be able to tell you when the circuit will be installed as he may not know himself. All too often the telephone company resists making appointments and turns up on-site unexpectedly. So who should you complain to? The VANS supplier or the national telephone company? The VANS supplier, in my opinion—for as part of your service from him you can expect to be relieved of the headache of managing difficult national telephone companies. However, do not expect too much—if you have previously worked with the telephone company you will know the problems he is facing.

Where the VANS offerings are strong is in their handling of international service difficulties. If there is a problem in the network there is no doubt that the VANS operator should be responsible for resolving it. Unlike their national telephone company competitors they cannot blame it on the national telephone company at the other end of the international connection. Unlike their national network competitors, most VANS have centralised management centres for quickly diagnosing and rectifying problems on an international scale. Some even offer their customers direct network management control over their use of the network. And as assurance of their service values, most of the VANS operators offer financial rebates where service levels do not meet those contracted. This also distinguishes them from many monopoly or ex-monopoly carriers.

Where most of the international VANS offerings are weak is in their coverage of locations. While many of them proudly proclaim how many countries they are operational in, this may mean no more than a single exchange somewhere in the capital city. In this case the cost of connecting individual customer sites (by leased circuit) is likely to be very high.

If the customer-site connection cost is passed straight on to the customer, he may face paying a considerable premium over and above the price of an equivalent international network service from the national telephone company. He will need to judge whether the extra service he receives from the VAN is worthwhile. The more aggressive operators subsidise the initial cost of customer connection lines as a way of developing business more quickly. They know that if they can build enough customers in a country quickly they can justify further exchanges in the more distant regions—so shortening customer connection distances and correspondingly reducing their costs.

What is behind the move towards international network service offerings? Simply the desire of the largest network companies to expand their international business and customer base. They fear erosion of their home markets through competition and see the new market opportunities in international networking. The small scale international VANS of today are only the first step in a period of change in the international telecom market that will shake up the world's network suppliers and create a few enormous multinational ones. But this will happen only at the pace that regulation and capital investment permits.

CULTURAL DIFFERENCES

Cultural differences affect both the users and suppliers of telecommunications services from country to country. What it may be acceptable to say in some countries, it might be important to write in others. Where electronic mail may be a standard way of doing business in one country, fax may be the best solution for another. In some countries it may be impolite not to answer within three rings, while in others the waiting might be a test of your real level of interest and urgency. Don't be put off from standardisation of methods—just be mindful of the different means that may be necessary to achieve it.

Dealing with telecom suppliers in advanced competitive markets is straightforward. You can ask for what you want and expect to get it. Price negotiation can be verging on the aggressive. Such style is perfect for the United States. In other countries, the market is less advanced and softer methods have to be used, sometimes accepting second best now in return for promise of improvement in the future. There is no point in merely upsetting the only supplier that you are able to buy service from.

Renowned poor performance of the telephone companies in some countries has created an acceptable excuse for poor performance of company telecom networks in these countries. This obviously makes the market prime for external competition but, for the local telecom manager, transferring to a more reliable supplier might present a risk of exposing his or his company's own management inadequacies (which might previously have been blamed on poor communications). This matter may also have to be dealt with.

PRODUCT AND SERVICE DIFFERENCES

The products and services of the various national telephone companies can differ greatly. In some countries nearly all calls go through first time, while in other developing regions you may have to book a call via the operator days in advance. The inevitable result is that the international services between countries comprise only the common capabilities of the two ends. Usually these will be substantially inferior to either of the two national services.

An example of the inferiority can be found even in a comparatively simple network service—leased circuits. In the United Kingdom it is possible to subscribe to a leased line service which allows customer adjustment of line capacity. A 2 Mbit/s line is provided between the customer's premises and the nearest telephone company site. The customer may control (within the scope of other similarly equipped locations) allocation of capacity at any specific time of day—for example, perhaps 8×256 kbit/s lines at night and 30×64 kbit/s lines during the day. In Germany, by contrast, the Deutsche Bundespost Telekom does not offer this service. However, its standard leased circuit offering, a DDV (Daten Direkt Verbindung) is one not normally matched in the UK. A DDV is always engineered with two permanently available paths through the network with automatic changeover, ensuring very high availability. You might like both the British and the German network features to be available internationally. Neither are. What are available are 'straight forward' full-time point-to-point circuits.

Those international services which do exist are described in specifications written by the CCITT (International Telephone and Telegraph Consultative Committee—a part of the International Telecommunications Union, ITU). The importance of the CCITT recommendations is such that all telecom equipment manufacturers seek to conform to them. Thus, for example, modems may be quoted to conform to V series recommendations, eg V22, V24, V32, V42 and/or V42*bis*, while international private leased circuits might conform to M1020, M1040, M1025 or G703.

In this way the CCITT has set the standard for basic network interconnection and so helped the cause of multilateral cooperation between national telephone companies the world over. Certainly many countries would remain cut off from the rest of the world had it not been for the work of this body. It is important that telecom managers developing international networks are familiar with the relevant CCITT recommendations.

Unfortunately, at a more advanced level of technology, the CCITT system has failed to move fast enough. The specifications of newer technologies have been a battle ground for international pride and commercial gain—too important a battle to be lost. Thus when CCITT participants have been unable to agree, diplomacy has seen to it that some recommendations include a number of different (and often incompatible) technical *options*. Even though two equipments may claim to conform to the same CCITT recommendation, they may not work with one another in practice! By the time the mess is sorted out, the recommendation is sometimes obsolescent (as for example those specifying a new generation of Telex, to be called Teletex, which failed even before products became available).

CCITT Recommendations used to be published only once every four years after substantial multilateral discussion and agreement. In an attempt to speed the process, CCITT has now moved to a system of publishing standards as they are completed. The continuation of the four-year study period, meanwhile, helps to ensure that the study groups to come up with some level of agreement and progress after four years.

There have been recent attempts among small groups of different national telephone companies and manufacturers to reach more quickly the commercial and technical agreements necessary for the introduction of new and advanced international network services. Pressured by increasing competition at home, like-minded companies have got together in an attempt to develop unique international service offerings. Some of these initiatives have helped to overcome the inevitable teething difficulties of new technology, and have speeded development a little.

One particular service developed in this way, called *one-stop shopping* or 'account management plus' has been quickly adopted by a number of national telephone companies. The service recognises the difficulty which normal international leased line customers have in ordering, administering and managing their circuits:

(1) Each circuit must be ordered twice — as a half circuit in both terminating countries.

(2) The customer may be asked to pay for one half on its delivery, even though it may be unusable because the other half has not been delivered yet.

(3) When he encounters a fault, the customer may have to make sure that he has reported it at the right end (ie the end where the fault is).

(4) He will receive two separate invoices in two separate currencies.

(5) Any purchasing discount may apply only to one half of the circuit.

The one-stop shopping service aims to address these difficulties by offering the customer a single point of order for his circuit, a single point of fault reporting and a single point of billing. Unfortunately the service has not been achieved by any radical improvement in cooperation between the national telephone companies. This would have been a welcome step forward. In practice the telephone company taking the now full circuit order merely transcribes this to the two normal half circuit order forms and despatches one set of forms to the distant carrier. Currently, I therefore prefer not to use such services widely, and instead like to retain a relationship with both end carriers. This ensures that should I have either operational or commercial problems to be resolved, there is some level of personal rapport available at both ends on which I can fall back.

TECHNOLOGICAL DIFFERENCES

The technology differences between countries have arisen largely because of the rigid independence of each of the national telephone companies as they develop and manufacture their own equipment. Even telephone sockets, which in most countries have appeared in their current form only over the last ten years, have been developed in glorious variety.

In Germany, the normal telephone socket has a specially developed electrical circuit associated with it to prevent the customer from plugging in

Figure 7.2   Technology differences

more than one telephone at a time—even if he has more than one socket—unless he has paid for multiple handset service as well as multiple socket service. The thinking goes, presumably, that:

(1) Extra electrical current is drawn from the line to feed the extra handsets and this should be charged for.

(2) There is good money to be made from charging customers for multiple handset service as well as for multiple socket service (even if they bought the handsets themselves).

(3) It is only Telekom staff who for the foreseeable future can be trusted to wire extra telephone sockets (and remember to wire-in the switch device).

What a contrast to the United States, where in some homes there is a phone in every room! What the American telephone companies have learned is that users make more calls if the phone is nearer to hand. The extra call revenue far outweighs any extra costs. In consequence there is a lot more freedom for self wiring of in-premises telephone cables.

One of the other notable technical differences in international telecom services is the different line-speed rates used on digital line systems. In Western Europe, the basic digital line is 64 kbit/s, and this is multiplexed to higher rates of 2 Mbit/s (so-called E1 rate), then to 8 Mbit/s, 34 Mbit/s, 140 Mbit/s and 560 Mbit/s. By contrast, in North America and the Pacific Rim, the basic 64 kbit/s *tributaries* are multiplexed up to 1.5 Mbit/s (so-called T1), then to 6 Mbit/s, 45 Mbit/s and 140 Mbit/s.

The two digital hierarchies are completely incompatible without the use of conversion equipment (even for single voice circuits at the 64 kbit/s or data circuits employing only 56 kbit/s of the available channel). The conversion equipment may be provided by the telephone company on international lines at rates below 1.5 Mbit/s, but for higher rates (ie 1.5 Mbit/s, 2 Mbit/s and above) you will almost certainly need to check that the equipment in both countries is capable of being connected to the line. Remember that 2 Mbit/s lines are 'non-standard' as far as North America and the Pacific Rim is concerned, while 1.5 Mbit/s is 'non-standard' in Europe.

Equipment or type approval can also vary from country to country and cause restriction. In many countries, equipment such as telephones, answering machines, office telephone systems, facsimile machines, etc cannot legally be plugged in to the public telephone socket unless they have been type approved. This means they have passed through a validation testing procedure which has proven that in every regard they meet the appropriate operational and safety standards.

Unfortunately, just because equipment is type approved in one country does not mean that it is in another. So even though it may work, and even though it may have a 'green dot' or an 'approved' label for one country it may not be legally used elsewhere. The most annoying part of this is the fact that the process of gaining equipment approval can sometimes be extremely lengthy for the manufacturer—taking sometimes six months or more. In the European Community 'harmonisation' work and new legislation should reduce problems of this sort between member countries, but to avoid such problems on a wider international scale, always ask the manufacturer to produce copies of the equipment approval certificates for all intended countries of operation before committing yourself to a large order of telecom equipment for international use.

The technical differences between national networks need not become a great barrier to the development of an international network, provided equipment suitable for each of the local markets is used. Using one of the multinational equipment manufacturers (Northern Telecom, AT&T, Philips, Siemens, Alcatel, Ericsson), quite sophisticated international telecom services can be developed without fear of local technical problems—by using the local variant of a common equipment type in each country. Most of these suppliers are willing to give a price discount on a multiple order, even though the delivery may be through different subsidiaries or distributors.

PRICE VARIATION

International prices for telecom network services vary widely and inexplicably. Sometimes it is three times as expensive for A to call B than it would be for B to call A. Sometimes even the structure of prices may evade understanding.

Unfortunately, public network service providers seem to revel in confusing their customers with extremely long and complicated price lists. Often, it is hard for a potential customer to understand what price he will have to pay for a given service without asking for a price quotation.

Many telephone companies, for example, continue to have a price structure for leased circuits which depends not on the distance between the two customer sites but on the distance between the two telephone exchanges which are nearest to these sites. Undoubtedly, the logic is that for the telephone company there is a *local loop* circuit cost at each end to connect the customer site to the exchange and to this is added the main circuit cost between the exchanges which is proportional to distance. Perfect logic, but it ignores the customer. How does he know where the nearest exchange is in order to calculate his costs, and anyway why should he have to know? He would not have expected to pay a furniture removals company based upon the distance between depots!

Worse still, some telephone companies charge for renting leaselines and have the further gall to charge volume dependent usage charges as well. Presumably, this is a ploy to protect the over-inflated profit margins on normal telephone and other switched network services—otherwise traffic might migrate to leaselines and yield far less revenue!

Always make sure you understand all the cost elements of a particular service. We explained in Chapter 4 about some of the hidden charges on public data networks (look out for usage volume charges, call set-up charges, call duration charges and all manner of surcharges applying simultaneously). Also understand any taxes which will be added. These can be a surprise too, particularly if you left 20% out of the costs of the calculation of total network costs!

To a large extent, the prices charged by telephone companies for international calls and leased circuits are governed by the accounting structure and accounting rates agreed under the auspices of CCITT. The accounting rates so defined determine the settlement payments which should be paid by one network operator to the other for his part in completing the call or for 'terminating' the half leased circuit. The payments are made by the network operator who collects the money from the customer, ie the 'originating' network owner. They recompense both 'terminating' network operator and also any 'transit' network operators (ie intermediate networks). Unfortunately, the CCITT recommendations on accounting and the out-of-date rates

Figure 7.3   Pricing variations

have become problematic in recent years, tending to grossly distort customer prices for international network services.

The problem is that the accounting rate recommendations have not kept pace with the radical shift in cost structure of international networks which has taken place in recent years. While they might once have been a fair means for dividing the revenue received from a particular service, they no longer are. The accounting rates are far too high to fairly reflect costs, and have distorted prices. The distortion is worst in cases where there is not a rough balance of incoming and outgoing traffic between two countries. Where there is balance of traffic, the accounting rate distortion is neutralised by the net null payment between operators.

The CCITT accounting rate recommendations are the main reason for the wide discrepancies in international telephone charges. Fortunately, there is a review currently underway, brought about by pressure from governments, telephone companies and customers alike to see prices brought into line with costs. Users can expect this review to have a reducing effect on their international phone bills. The per kilometre charge ought to be reduced in line with long distance national telephone calls. This should make a big difference in continental regions—such as the European Community, where today a 100 km call between France and Germany (or other neighbouring countries) is charged at an international tariff far above the long distance rates for equivalent calls made nationally.

In the meantime, most companies will have to continue to pay the rates as they are. That is, unless they themselves are big enough to arbitrage their international telecom usage. In the past there have been specialist arbitrage businesses set up, for example telex refiling agencies. Typically, a telex refiling agency would have been situated in London. It would have offered to forward telexes received from the European continent for delivery in the United States. Why would it have been attractive? Because transatlantic rates from the United Kingdom were far cheaper than from other continental European countries.

Today's arbitrage method is the hub network. This is a private network in a star-shaped topology centred in a particular country because of the low international leased circuit or telephone charges in that country. Historical hub countries have been the United States, the United Kingdom for Europe, and Singapore and Hong Kong for the Pacific Rim region.

An alternative arbitrage method is the use of a specialist international telephone reseller (there are several based in the United States). You call the reseller. His equipment calls you back automatically and also connects to your destination. Both calls are made from the United States where international rates are very low.

8

Communicating with Confidence

Improvements in, and expansions of, communications systems and networks have left many companies open to breaches in confidentiality, industrial espionage and abuse. Sometimes such breaches go unnoticed for long periods, and can have serious business or cost implications. Equally damaging can be the impact of simple mistakes—misinterpreted, or distorted information. Increased belief in the reliability of systems and the accuracy of information has brought great gains in efficiency, but blind belief suppresses the questions which might have confirmed the need for corrections. This chapter describes the various levels of information protection which may be provided by different types of telecommunications networks, and the corresponding risks. It goes on to make practical suggestions about how a company's protection needs could be assessed, and how different types of information can best be secured in transit.

THE TRADE-OFF BETWEEN CONFIDENTIALITY AND INTERCONNECTIVITY

The man who sold the first telephone must have been a brilliant salesman— for there was no-one for the first customer to talk to! On the other hand, what confidence the customer could have had that there were no eavesdroppers on his conversations! The simplicity of the message should be a warning to all: the more people on your network, the greater your risk.

As the number of connections on a network increases, users are subjected to:

(1) The risk of interception, 'tapping' or eavesdropping.

(2) Greater uncertainty about who they are communicating with. (Have you

reached the right telephone or not? Which caller might be masquerading as someone else?)

(3) The risk of time-wasting mistakes. (An incorrect access to a database or a misinterpretation of data may lead to the corruption or deletion of substantial amounts of data.)

(4) The nuisance of disturbance. ('Wrong number' calls, unsolicited calls from salesmen, and, worse still, 'forced entry' by computer 'hackers', or abuse of the network by third parties to gain free calls at your expense.)

Too often, much thought goes into improving the connectivity of networks, but too little is applied to information protection. Risks creep in—often unnoticed. We discuss next the different types of protection which are available.

DIFFERENT TYPES OF PROTECTION

The information conveyed across communication networks may be protected from external distortion or abuse by any one of four basic means (see Figure 8.1):

(1) *Encryption*: coding of the information, so that only the desired sender and receiver of the information can understand it, and can tell if it has been distorted.

(2) *Network access control*: allowing only authorised users to gain access to the communications network at its entry point.

(3) *Path protection*: permitting only authorised users to use specific network paths.

(4) *Destination access control*: allowing only authorised users to exit the network on a specific line, or to gain access to a specific user.

A combination of the four different protection methods will give the maximum overall security. Methods which are available in the individual categories are described below:

Encryption

Encryption (sometimes called 'scrambling') is available for the protection of both speech and data information. A cypher or electronic 'algorithm' can be

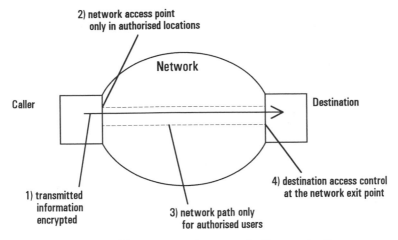

Figure 8.1   Four aspects of communications protection

used to code the information in such a way that it appears to third parties like meaningless garbage. A combination of a known codeword and a decoding formula are required at the receiving end to reconvert the message into something meaningful. The most sophisticated encryption devices were developed initially for military use. They continuously change the precise codewords and/or algorithms which are being used, and employ special means to detect possible disturbances and errors.

To give maximum protection, information encryption needs to be coded as near to the source and decoded as near to the destination as possible. There is nothing to compare with speaking a language which only you and your fellow communicator understand!

In a technical sense the earliest opportunity for encryption is the caller's handset. Sometimes, either for technical or economic reasons, this point is not feasible and the encryption is first carried out deeper in a telecommunication network. Thus, for example, a whole site might be protected with only a few encryption devices on the outgoing lines rather than equipping each PBX extension separately. Clearly the risks are then higher since the PBX itself could be 'tapped'—so a judgement of the protection necessary may need to be made.

For most commercial concerns I do not believe the security risks arising from technical interception of signals are that great. It is much simpler to overhear conversations on the train, read fax messages carelessly left on unattended fax machines or 'bug' someone's office than it is to intercept messages half way across a network.

For maximum protection of data, the data itself should always be stored in an encrypted form, and not just encrypted at times when it is to be carried across telecommunications networks. Permanent encryption of the data renders it in a meaningless or inaccessible form for even the most deter-mined computer hacker. Thus, for example, encrypted confidential informa-tion held on an executive's laptop computer can be prevented from falling into unwanted hands, should the laptop go missing.

Network Access Control

By controlling who has access to a network we minimise both intentional and unintentional disturbances to communication. In much the same way we might reduce the road hold-ups, hazards and hijacks by limiting the number of cars on the road.

The simplest way of limiting network access is to restrict the number of network connections. Without a connection, a third party cannot access a network and cannot cause disturbance. The physical security of connections which do exist (ie lock and key) may also be important for very high security needs.

Entry to a network can also be protected by password or equivalent software-based means. The simplest procedures require a user to 'log on' with a recognised username, and then further be able to provide a cor-responding authorisation code or *personal identification number* (PIN).

The problem with simple password access control methods is that people determined to get in just keep trying different combinations until they stumble on a valid password. Aided by computers, the first 'hackers' simply tried all the possible password combinations. The problem can be alleviated to some extent by limiting the number of attempts which may be made consecutively (bank cash teller machines, for example, typically retain the customer's card if he does not type in the correct authorisation code within three attempts).

More secure password control systems require the user first to produce some sort of physical token (eg a key or a magnetic card). Without the key or card the system simply does not allow other potential intruders to start trying passwords.

Path Protection

The communication path itself is bound to run through public places and in consequence past sources of potential eavesdropping, interception and dis-turbance. The best path protection depends upon the right combination of physical and electrical telecommunication techniques, but from the serious eavesdropper there is no absolute protection. Encryption, as already discussed, prevents the eavesdropper from understanding what he might pick up. To reduce the risk of interception the path should be kept as short as possible and not used if electrical disturbances are detected upon it. There is nothing better than sitting in the same room!

In the early days of telephony, individual wires were used for individual calls and thus the physical paths for all callers were separate. Laying a separate cable continues to be a means of security for some. Some firms, for example, order their 'own' point-to-point leased lines from remote sites to their computer centre to ensure that only authorised callers can access their data. But for the determined eavesdropper the physical separation may be an advantage—it is much easier to identify the right cable and tap into it at a manhole in the street. Alternatively, without tapping, he can surround a copper cable with a detection device to sense the electromagnetic signals passing along the cable, and interpret these for his own use.

The electromagnetic radiation emitted by high speed data cables is well recognised, and can lead companies to information corruptions and losses of both malicious and non-malicious nature. Many high speed data cables laid alongside one another tend to cause crosstalk disturbances for speech and corruption problems for data. High power lines laid alongside communication cables also can cause problems. The correct name for this effect is Electromagnetic interference (EMI). EMI has been a particular problem is recent years for companies running high speed LAN systems and for PC users who also have portable telephones. Careful cable route planning (including rigid observance of equipment operating conditions and specified maximum workable cable lengths) is therefore very important.

Alternatively, since optical fibre cable is relatively emission-free, it can provide protection both against eavesdropping and EMI problems.

Where radio is used as the communications path (you may not know this if you order a leased line from the telephone company), interception by eavesdroppers is made even simpler. Protection of radio (both from radio interference and from eavesdropping) can be achieved at least to some extent by new methods such as frequency hopping. In this method both transmitter and receiver jump in synchronism (every few fractions of a second) between different carrier frequencies. Jumping about like this reduces the possible chance of prolonged interference which may be present on a particular frequency, and makes it very difficult for eavesdroppers to catch much of a conversation.

Most modern telecommunications devices use electrical methods to enable many different communications to coexist on the same physical cable at the same time. On the one hand this makes it harder to perform interception through tapping since the electrical signal carried by the wire has to be decomposed into its constituent parts before any sense can be made of a particular communication. On the other hand, it may mean that an electrically coded version of your information is available in the machine of someone you might like to keep it from. A message sent across a LAN, for example, may appear to go directly from one PC to another. In reality the message is broadcast to all PCs connected to the LAN and the LAN software is designed to ensure that only the intended recipient PC is activated to decode it.

In practice, path protection across LANs is not possible. If such paths cannot be avoided by sensitive data transmissions then data encryption must be used.

Destination Access Control

Protection applied at the destination end is analogous to the keep of a medieval castle—having got past the other layers of protection it is the last hope of preventing a raider from looting your prized possessions.

On highly interconnected access networks, destination protection may be the only feasible means available for securing data resources which must be shared and used by different groups of people. Typically, companies apply access control methods at a computer centre entry point, for example. A much used protection method is a simple password authorisation within the computer application software, but the level of security can be substantially improved by combining this with one of two types of feature which may be offered within the feeder network—either Calling Line Identity (CLI) or Closed User Group (CUG).

Calling Line Identity (CLI) is a feature available on telex networks, on X25 packet switched data networks and on modern ISDN telephone networks. The *network itself* identifies the caller to the receiver, thus giving the receiver the opportunity to refuse the call if it is from an unauthorised calling location (see Figure 8.2). Call-in to a company's computer centre can thus be restricted to remote company locations. Password protection should additionally be applied as a safeguard against intruders in these sites.

Not all systems which might appear to offer calling line identity (CLI, already discussed in Chapter 3) are reliable. Fax machines, for example, often letterhead their messages with 'sent from' and 'sent to' telephone numbers. These are unreliable. They are only numbers which the machine owner has programmed in himself. It is thus very easy for the would-be criminal to masquerade under another telephone number (either as caller or as receiver) to send false information or obtain confidential papers. Even though you may have dialled a given telephone number correctly, you have no idea where you may have been automatically diverted to!

Closed User Group (CUG) facility is common in data networks. To a given exit connection from the network for which a CUG has been defined, only pre-determined calling connections (as determined by the network

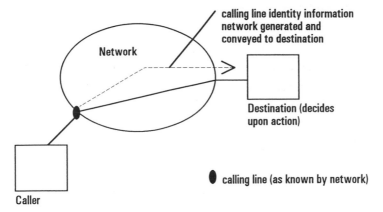

Figure 8.2   Calling line identity

itself) are permitted to make calls. Typically a small number of connections within a CUG are permitted to call one another. Additionally, they may be able to *call* users outside the CUG, but these general users will not be able to call back. In effect, communication to a member of the group is closed except for the other members of the group, hence the name. The principles of CUG are illustrated in Figure 8.3.

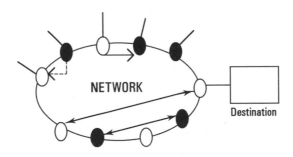

○ Member connections of the Closed User Group-can call to all other 'white' or 'black' connections

● Ordinary network connections - can only call other 'black' connections

↗ Calls possible between these endpoints in either direction

→ Calls possible only when set up in the specified direction

--→ Calls not possible to be set up in this direction between these endpoints

Figure 8.3   The principle of closed user groups (CUGSs)

SPECIFIC TECHNICAL RISKS

What are the main technical risks leading to potential network abuse, breaches in confidentiality or simple corruption of information? What can be done to avoid them?

Carelessness

Always check addresses. I was once amazed to receive some UK-government classified SECRET documents that should have been sent to one of my namesakes!

Why even think about encrypting a fax message between sending and receiving machines, if either machine is to be left unattended? Do not contemplate reading it on the train or talking about it on the bus.

Computer system passwords should be changed regularly. If possible, password software should be written so that it demands a regular change of password, does not allow users to use their own names, and does not allow any previously used passwords to be re-used.

Ex-employees should be denied access to computer systems and databanks by changing system passwords and by cancelling any personal user accounts.

Computer systems designed to restrict write-access to a limited number of authorised users are less liable to be corrupted by simple errors. Holding the company's entire customer records in a PC-based spreadsheet software leaves it very prone to unintentional corruption or deletion by occasional users of the data. Any changes to a database should first be confirmed by the user (eg 'update database with 25 new records? Confirm or cancel'). Subsequently, the system software should perform certain plausibility checks before the old data is replaced (eg can a person claiming social security really have been born in 1870?).

Ensuring proper and regular back-ups of computer data helps guard against corruption or loss due to viruses, intruders, technical failures or simple mistakes. Daily or weekly back-ups should be archived 'off-line'.

Simple precautions properly applied would dramatically reduce the risk of most commercial concerns!

Call Records

For some very senstitive commercial issues, say when contemplating a company takeover, it may be important to a senior company executive that no-one should know he is even in contact with a particular company or advisor. Such company executives should be reminded of the increasing commonality of itemised call records from telephone companies, and similar 'call logging' records which can be derived from in-house office telephone systems. Such devices keep a record of the telephone numbers called by each telephone line extension.

Mimicked Identity

Sometimes information can be gained under false pretences by claiming to be someone authorised to receive that information. Just as problematic and probably easier, false information could be fed into an organisation or system to confuse or corrupt it. Virus software, for example, once into a computer can wreak almost unlimited damage.

Do not accept identity information which you cannot trust (for example, the 'sent from' or 'sent to' telephone numbers which appear on fax messages. If the identity of a caller or destination should be confirmed use a network technology which can be relied upon to confirm addresses (eg Telex).

Do not forget the possibility of a called destination having been diverted to somewhere else. Modern telephone networks give householders, for example, the chance to divert calls to their holiday cottage while on vacation. They also provide a new opportunity for criminals. A telex network answerback confirms the right destination has been reached, and similar called line identity (the destination number) can provide assurance on X25, ISDN and other modern networks.

Radio Transmission, LANs and other Broadcast-type Media

Broadcast-type telecommunications media, while otherwise being technically very reliable, are not well-suited to high security applications. The Princess of Wales (Diana) discovered to her cost just how easily mobile telephones (cellular radio handsets) can be intercepted. But other broadcast telecommunication media may not be so apparent to users—satellite, LANs and radio sections of 'leased lines' rented from the telephone company.

Satellite transmission has proved to be one of the most reliable means of international telecommunication. Satellite media do not suffer the disturbances of cables by fishing trawlers and by sharks and achieve near 100% availability over long periods of time. But from a security standpoint, just about anyone can pick up a satellite signal. Thus satellite pay-TV channels need much more sophisticated coding equipment than do cable TV stations to prevent unauthorised viewing.

Local area networks of interconnected PCs work by broadcasting information across themselves. So while LANs achieve a very high degree of connectivity (particularly those connected to the public *Internet* network), they could also present a security risk for sensitive information.

EMI (Electromagnetic Interference)

Electromagnetic interference has recently become a significant problem as the result of high power and high speed data communications devices (eg mobile telephones and office LAN systems). EMI can lead to corruption of data information and general line degradation, particularly with intermittent and unpredictable errors.

The problem of EMI is recognised as so acute that a range of international technical conformance standards has been developed which define the acceptable electromagnetic radiation of individual devices. In practical office communication terms, the most common problems are experienced with high speed data networks (eg LANs), particularly when the cabling has not been well designed. Simple precautions are:

- The rigid separation of telecommunications and power cabling in office buildings.

- The use of specified cable material only.

- The rigid observance of specified maximum cable lengths.

Message Switching Networks

Certain telecommunications networks (eg electronic mail networks, voice-mail networks, some fax machines and fax networks and X400 networks) carry whole messages in a store-and-forward fashion. The sender creates the message and posts it into the network, where it is stored in its entirety. The message subsequently progresses step-wise across the network as the availability of resources permit. Either the message will be automatically delivered to the user (eg fax) or it may wait for him to pick it up (eg electronic mail).

Message switching networks offer their users a higher level of confidence that messages will be delivered correctly and completely, and usually can give confirmation of receipt. At one level, modern message systems (eg electronic mail or voicemail) ensure that messages are read or heard by a manager himself rather than by his secretary. But for very highly confidential information, users need to take into account the fact that a complete copy of the message is stored somewhere in the transmitting network.

'Deletion' of a message from your mailbox may prevent you as a user from further accessing a message, but should not be taken to imply that the information itself has been obliterated from its storage place. A technical specialist with the right access may still be able to retrieve it.

Public telecommunication carriers in most countries are obliged by law to ensure absolute confidentiality of transmitted information and proper deletion once the transmission is completed successfully. I believe this legal protection is adequate for the confidentiality needs of most commercial concerns, although for matters of national security it will not be.

Some modern fax machines (particularly those which offer 'broadcast' facility) also work by first storing electronically the information making up the fax. It may thus be possible for others to retrieve your message from the sending machine, even though you have removed the original paper copy.

Other Types of Network Abuse

Finally in this section, let me point out that one of the most common motivations for network intrusions is the simple criminal desire to get something for nothing—perhaps telephone calls at your expense.

One of the easiest ways to create this opportunity for an outsider is to set up a network with both dial-on and dial-off capability. Thus some companies, for example, give a reverse-charge network dial-on capability to enable their executives to access their electronic mailboxes from home without expense. Some networks simultaneously offer a dial-off facility. Thus, for example, the London office of a company might call anywhere in the United States for domestic tariff, by first using a leased line to the company's New York office, and then 'dialling-off' into the local US telephone company. The outsider can make all the calls he wants—entirely at company expense—unless the network is well enough designed to prevent simultaneous dial-on and dial-off by the same call (see Figure 8.4).

Alternatively, dial-back can be used instead of dial-on. Dial-back similarly reverses the charges for the caller (other than the cost of the initial set-up call), but in addition enables the company to have greater confidence that only authorised callers (ie known telephone numbers) are originating calls.

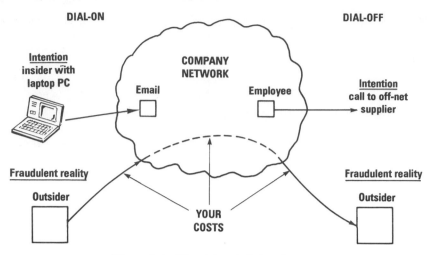

Figure 8.4   The risks of dial-on/dial-off

THE RIGHT CONFIDENTIALITY POLICY FOR COMMUNICATION

Keep it simple! Minimise the number of potential weaknesses by standardising on accepted company telecommunications media (eg electronic mail, telephone) and building adequate safeguards around these 'normal' methods. Make sure that the company's basic security and confidentiality policy covers the particulars of acceptable telecommunications media for each classification of document or other information.

The human links in communication are usually the weakest and the most prone to 'leak' information, so ensure that adequate discipline is applied to maintaining professional care. Automate processes like the changing of computer passwords to ensure they are updated regularly.

As to specific DON'Ts: don't use fax or mobile phones for company information classified as 'confidential'.

AN ONUS ON COMMUNICATORS

Confidence in communication—the reliance on the safe delivery of accurate information into the right hands—depends most on the right choice of communications media by the originating party and professional care by both parties.

Only the originator can be to blame if he sends a confidential message to an unattended fax machine. But the call receiver also can give important information away. I have been amazed to observe investigative journalists at work—simply calling up company representatives during periods of intense company activity (eg at a time of rumoured merger) and asking speculative questions. The journalist may know nothing to start with, but direct questions act like a bait to an unsuspecting manager, and may draw interesting information if they strike close to a current 'truth'.

I have also been amazed at what I have overheard on buses!

9
Some Final Thoughts

To survive, businesses need to know how to keep themselves on the move—in step with the market they are serving—and innovative enough to keep themselves growing. A company telecom department faces no less a challenge. To be successful, it needs to be adapting continually to new management and technology ideas, taking on those which promise cost savings or wider business innovation.

It is impossible to write a step-by-step handbook of communications innovation, but this chapter does cover a few systematic means by which information can be collected about new and emerging possibilities. We then conclude with a few specific ideas on the use of today's 'leading edge' technologies. Ideas, after all, are the fuel of innovation. Ideas lead to further ideas. . ..

WHERE DO YOU GET NEW IDEAS?

Everywhere. You can get ideas by observing others and adapting their ideas, or simply by copying them. Very few ideas are revolutionary in their own right—nearly all are developed from the adaptation of previous ideas. Companies in one industry may leap ahead of their competitors simply by importing new ideas from other industries or by using old ideas on new technologies.

Go to exhibitions and seminars. Listen to others. Continuously poll your suppliers and all passing salesmen, for you cannot expect to have ideas in isolation. Ideas have to be cultivated. The individual with the best idea doesn't necessarily know how best to apply it. The laser, for example, had to wait many years before optical fibre cables were developed to facilitate a telecommunications application.

Presenting new technologies to potential users amongst other business departments helps to stimulate general business folk to new ideas about how to apply new communications techniques. These are the most valuable of ideas, since not only do they take the business further forward, they also enjoy its greatest commitment in ensuring realisation.

Keep returning to basic business goals when assessing new technologies—"How can we get closer to customers? How can we reduce the cost of processing an order?". This helps to focus attention on the aspects of a technology which are most valuable.

THE OPPORTUNITY IN NEW TECHNOLOGY

It is new technology which offers the best opportunity for improvement. Developments based on older technologies tend to bring cost reductions or service improvements, but new developments bring the leap-forwards.

Knowing how best to employ new technology is rather more difficult, for it is not always immediately apparent what the full scale of possibilities might be. This should not be allowed to put you off from a review, for the harnessing of new technology could potentially revolutionise your company business, bringing significant competitive advantage. The advent of small dish satellite technology has, for example, brought a new era to the world of television. Now satellite television stations can reach audiences spread over whole continents, and draw customers for their 'movie channels' from a much larger potential customer base.

How do you determine how a new technology might add to your business? How, for example, might mobile communications best be employed within your business? What scope is there for the use of satellite communications? What other technologies should you be considering? The answer to all these questions may be that no new technology has any appropriate application at the moment, but without review you will not be able to discount anything.

The review can only be conducted in conjunction with the new technology supplier, for his intimate knowledge of the technology and its capabilities is critical to the development of ideas. Simultaneously, your injection of current business problems and ideas for business development is just as important; you and your company need to educate your suppliers about your business and get them to start doing some of the thinking on your behalf—presenting their product in a way that has business significance for you. They may have to adapt their product, but provided it is worth while to do so, any good supplier will be looking for such opportunity, for it is a new market opportunity for them.

The process should be an iterative one, with continuing injection of ideas from both you and your supplier. But the results will not come in a steady, project-managed flow. It may take many months or even years of discussion, on and off, sometimes intense, sometimes not; sometimes long gaps. Don't be put off. There will be breakthroughs and let downs—on balance probably mostly let downs, for you have to sift through a lot of ideas before you find not only a good one, but one which can be moulded into a viable business proposition.

THE IMPORTANCE OF ANTICIPATION

The long term payback periods of telecom equipment have a dampening effect on the rate at which systems can be replaced with latest technology. This in turn means that a telecom manager may, at any point in time, be lacking the right technology to achieve maximum benefit and economy of his network. The only way to avoid this problem is to try to anticipate trends in the market. Planning is not about meeting yesterday's needs with today's technology. Rather it is about the future—getting the right technology and networks in place for the moment *before* you need them. You can either guess what is coming or you can shrewdly get on a few inside tracks with the suppliers.

TECHNOLOGY WITH PROSPECTS

What are the emerging technologies that will shape business telecommunic- ations over the next ten years? In my view, they are the following:

- electronic mail and electronic data interchange (EDI);
- mobile voice and data networks;
- telephone businesses, 'distance learning' and 'homeworking';
- 'groupware' software tools and group communications systems;
- video and graphic transfer systems;
- voice processing and recognition, translation and 'expert' diagnosis sys- tems;
- higher speed and faster response 'backbone' network technologies.

For many company executives the laptop is already part of their everyday routine, as is the daily log-in to their electronic mailbox. Electronic mail between companies (so-called electronic data interchange), however, remains in its early stages. Though a relatively high number of companies may already be using EDI in some form for certain orders or money transfers, few have yet realised even half of the potential automation. As the incompatibility of different systems is broken down by EDI technology standards (X400, X500 and EDIFACT in particular), this revolution will continue.

For senior executives, the mobile phone is also part of the routine. Within the next 10 years I expect a wider migration to mobile network technologies. Cheaper costs of networks and handsets will bring mobile telephones to a much wider business and private clientele. Businesses will thus become more responsive as the result of (i) making sales and field staff more accessible, and (ii) freeing more managerial staff to 'roam' from their desks. Meanwhile, mobile data networks will offer new opportunities such as the ability of carriage and haulage companies to track their goods manifests and deliveries.

As confidence in, and acceptance of 'business by telecommunication' increases, new business practices will emerge. Citibank of the United States and Midland Bank of the UK have both already proved the scope for taking on high street banks with a combination of streetside cash dispensing machines and telephone advice services. No more is there the potential of a face-to-face talk with the local bank manager—costs have been much reduced by running with far fewer, but specialist staff.

Distance learning is the concept of running courses by video or other communication links between teacher and distributed students. Homeworking is a concept in which companies provide employees with sophisticated telecommunications equipment in order to allow them to work permanently from their own homes without loss of productivity.

Groupware is the term applied to a new breed of software—of a type particularly designed to allow groups of people simultaneously to work together. Thus, for example, several different users on a LAN system might simultaneously develop a document or a database. Groupware enables them to keep up-to-date with all the changes other users are making—as they are making them. These systems are most likely to be developed for PCs or UNIX-based computers on LANs. They will require fast and substantial data transfer between systems, and I thus expect a further boom in demand for LAN-to-LAN telecommunication traffic, using technologies like frame relay and TCP/IP. An example of an existing groupware software is 'Lotus Notes'.

Live video transmission today remains expensive due to the relatively high bandwidth required for it. However, there have been significant advances in technology which have already reduced the unit costs of terminal equipment and the bandwidth required during usage. Demand for this technology is set to explode—since the price/value balance is nearly right and the companies who have invested in videotelecommunications development are already keen to reap their rewards.

I expect videoconferencing and picture telephones to become a substitute for some travel but also an enabler of new meetings which might otherwise not have taken place. In parallel, I expect a number of new 'image' applications to appear. Some companies, for example, already keep a video 'scan' of important plans and other documents, enabling them to be available 'on-line' to a large cross-section of their staff at short notice. Thus a surveyor in a property company might be able to refer to building plans from on-site. Previous photographs held in a picture store might also help him assess any damage to building or interior.

Voice processing is already common in North America as a means of enabling telephone callers to question an information source. The next step will be more sophisticated systems with voice recognition, language translation and so-called 'expert' computer systems which give advice based on their previous actions and experiences. The first steps towards allowing a man to talk intelligently to a machine have already been taken!

As for 'backbone' communications technologies—for use as the central core of a corporate telecommunications network—the trend in demand for higher speeds, faster response times and greater flexibility will continue. The most important technology currently to watch is ATM (asynchronous transfer mode). ATM looks likely to be the first truly flexible technology—able to support voice, data and video equally optimally. Early ATM devices are available, but I believe the technology needs a further few years development before it will be really ready.

WHAT TO DO WITH YOUR EXISTING TECHNOLOGY

Keep it as long as possible. Once you understand the new potentials of older technologies underlying your existing networks, it is usually obvious (and sometimes surprising) what further benefits can be achieved, and is easy to assess the costs of a 'marginal' network development. These are important to establish since the capital investment in telecommunications networks is substantial, and existing technology cannot be lightly thrown away.

The main hurdle is usually the achievement of a full understanding of how far the technology can be developed and what is possible as a 'marginal' development of an existing network. What new software or hardware developments to the base hardware can be added to the network at minimal cost? What extra 'peripheral' equipment could be added to the network to bring new capabilities (eg voicemail hardware added to an office telephone system)? Press your supplier for a continuous flow of ideas and product updates—make him anticipate and plan for your future needs.

NEW BUSINESS POSSIBILITIES

Finally, a short lesson in lateral thinking. No, not one of those infernal party games where you have to explain some most absurd situation:

Challenge. *Explain why a man, who turns up at a hotel pushing his car, knows instantly that he is bankrupt.*

Answer. *The man is playing the board game 'Monopoly', which is based on the theme of property development. One of the player's pieces is a small model car. Should this player shake the dice and move his 'car' on to one of the board squares (city streets) where his opponent has erected a hotel, he will almost certainly lose all his money, through the 'rent' that he becomes due to pay.*

No, the challenge of telecommunications is much harder. Number one, it is for real. Number two, there is no right answer. Number three, there is no-one to lead you through it.

The challenge is to find a radical new business opportunity out of the combination of your company's current business strengths, its assets and the capabilities of telecommunications technology.

Now you are on your own. Good luck!!

Glossary

CCITT (International Telephone & Telegraph Consultative Committee)

An influential subcommittee of the International Telecommunications Union (ITU), which in itself is an agency of the United Nations. CCITT is one of the leading bodies in setting technical and operational standards for international telecommunication networks.

CCITT Recommendation

A technical standard issued by the CCITT. They are categorised into a number of 'Recommendation series'. Individual recommendations are thus usually known by their series letter and a number; X25, for example, is a particular type of 'data communications network interface' in the X series of recommendations; in fact, the one used in packet-switched networks.

LAN (Local Area Network)

A local area network (LAN) is a special type of data network optimised for interconnecting personal computers, printers, 'file servers' and other similar devices within an office environment. The most common types are 'Ethernet LANs' and 'Token Ring LANs'

Private Network

The term applied to privately-owned internal company networks typically serving large campuses or linking different sites. Often, the switching equipment in such networks is privately owned and operated, but lines between sites are leased from the monopoly telephone company.

PTT (Post, Telegraph and Telephone organisation)

PTT is the general acronym used to describe the typically state-owned, monopoly, telephone network operator. Many of these organisations retain their responsibility for postal services as well.

Public Network

The term applied to any type of telephone, data or other telecommunication network type available for public subscription. The term is often used to apply to the networks and services offered by national monopoly telephone companies, but also applies to the services from other carriers in countries where network competition exists.

VANS (Value Added Network Service)

A term with particular legal significance in countries where the telecom network services market has already been liberalised. In many 'liberalised' countries a monopoly has been retained on 'basic' transmission (ie street cabling) and public telephone network services, but value added network services (VANS) operators may provide other more specialist services using the basic services as a foundation.

WAN (Wide Area Network)

A term applied to a telecommunication network interlinking sites which are widely geographically spread (across a country, a continent or even globally dispersed).

Index